UNIVERSAL MATHEMATICS

8 and 9

BY: ROLAND PIRES

WWW.UNIVERSALMATH.NET

Introduction

I have been teaching math for the past ten years and am using a new approach to many topics such as integers, fractions equations etc. My first topic in the book covers integers. In just a few pages, the addition, subtraction, multiplication and division of integers are explained. I do not use counters or tiles to explain integer operations, as these methods are not easily understood by students. Fraction addition, subtraction, multiplication and division are again explained easily without the use of grids or arrows which many students find difficult to understand.

This book will help you understand math which is the sole reason for writing this book. Understanding math is more important than trying to follow a particular method such as using counters or tiles to understand integer operations. In certain books, whole chapters are written using counters and tiles to explain integer operations. I have explained integers in a few easily understood pages.

A student asked me to teach her fraction multiplication and division using fraction strips. She said that nobody in the class could understand fraction strips. After explaining the method which was necessary for her to complete her homework, I explained to her that when multiplying two fractions, just multiply the two numerators to get your new numerator and multiply the denominators to get the new denominator. The new numerator and denominator is the answer. She was surprised.

The materials covered in this book are based on Grade **8** and **9** curriculum. Methods that are difficult to understand such as described above are not used. Use this book in conjunction with your regular text book. With the number of worked out examples and explanations it may not be necessary to hire a tutor.

This book will help you understand math with easy to understand explanations and methods. You will not find wording of problems that may be hard to understand because this is a math book and as such should be concentrated on math.

This book is written for Grade **8**, Grade **9** students and any adults who might want to learn math. To get the maximum benefit, no pages should be omitted, so as to understand later topics. You will not find yourself turning through pages that are useless. The topics are arranged in such a way that knowledge gained from previous chapters is necessary to understand later chapters. (Integers are taught before fractions, since the knowledge of integers is necessary to add and subtract fractions).

Grade **8** is covered in Chapters **1** to **9** and Grade **9**, in Chapters **10** to **16**. Students starting in Grade **9** should be familiar with integers, fractions, decimals, percent, rates, ratios and equations covered in Grade **8** section of this book. It is essential that Grade **9** students read and practice these topics so as to understand Grade **9**.

Mathematics when presented properly is not that complicated.

CONTENTS

CHAPTER 1

INTEGERS

CHAPTER 2

FRACTIONS DECIMALS AND PERCENT

CHAPTER 3

RATES AND RATIOS

CHAPTER 4

MEASUREMENTS

CHAPTER 5

EQUATIONS 1

CHAPTER 6

GEOMETRY 1

CHAPTER 7

BEDMAS

CHAPTER 8

TRANSFORMATIONS

CHAPTER 9

PROBABILITIES

CHAPTER 10

REVIEW OF ESSENTIAL SKILLS FOR GRADE 9

CHAPTER 11

EXPONENTS

CHAPTER 12

BEDMAS 2

CHAPTER 13

POLYNOMIALS

CHAPTER 14

EQUATIONS 2

CHAPTER 15

GEOMETRY 2

CHAPTER 16

EQUATIONS 3

CHAPTER 1

INTEGERS

An integer is a whole number and can be positive or negative: for example **+3, +6, -9, -34** are integers; **1.5, -4.5, 3.7** are not integers.

The sign in front of a number indicates whether the number is positive or negative. *If there is no sign in front of the number, then the number is positive; for example* **5** *means* **+5**.

The number line below shows positive and negative integers.

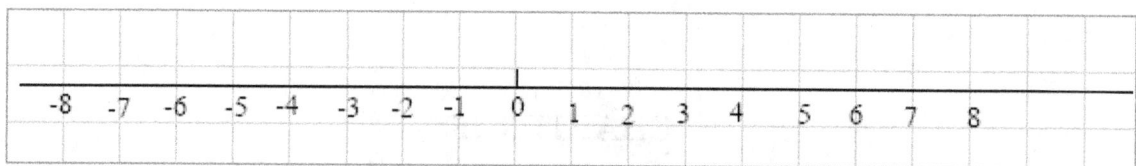

Integers to the left of zero are negative and integers to the right of zero are positive. As we move to the left on the number line, the integers get smaller and moving to the right, the integers get larger. **-8** is less than **(<) -7; -7** is greater than **(>) -8**.

Examples

$$-6 \ < \ -2$$
$$4 \ > \ 1$$
$$-5 \ < \ -2$$
$$-3 \ > \ -5$$

<u>*Rules for adding positive and negative integers*</u>

When the signs are the same, add the integers and the answer takes the sign of the integers.

<u>Adding negative integers</u>

Example

Adding **-5 and -6 = -11**

Because the signs are the same, add the integers **5 + 6 = 11**; the sign is negative because both integers are negative.

When adding any number of negative integers, the answer will always be negative.

Adding **-6, -5,-4** and **-3 = -18**   {The same **as (-6)+(-5)+(-4)+(-3)**} This second method is used by some.

Adding positive integers

Example

    Adding **5 + 5 = 10**

Because the signs are the same, add the integers **5 + 5 = 10**; the sign is positive because both integers are positive.

Notice that the first **5** does not have a plus sign. If the first integer is positive, the positive sign is omitted.

When adding any number of positive integers, the answer will always be positive

    9 + 8 + 5 = 22

In this book, integers will always be written in this form

    - 3 - 6 - 8 = - 17

Integers will not be written as follows: **(- 3) + (- 6) + (- 8) = - 17**. (This method is used in lower grades and has to be unlearned in the higher grades to **- 3 - 6 - 8 = - 17**).

In Grade **9**, the result of an equation may look something like this:
    x = - 4 - 9 - 8   but not   **x = (- 4) + (- 9) + (- 8)**
The answer is **x = - 21**

The reason for omitting the brackets and the sign between them in this book, is that they serve no useful purpose.

Given a number of negative integers such as **- 5 - 10 - 6 - 7**, there is no choice but to add them. The answer is **- 28**.

The sign in front of an integer indicates whether it is positive or negative, so it is not necessary to put brackets to indicate that the sign within the bracket belongs to the integer.

Adding positive and negative integers

When the signs are not the same, subtract and the answer takes the sign of the larger integer.

Example

    Adding **- 5** and **10 = + 5**

Since the signs are not the same, subtract **5** from **10** and the answer is **5**. The sign in the answer is positive, because **10** is positive and **10** is the larger than **5**.

Example

Adding **5** and **- 10 = - 5**

Because the signs are not the same, subtract **5** from **10** and the answer is **5,** but the sign in the answer is negative, because **10** is the larger number which has a minus in front of it; actually **- 10** is less than **+ 5**, but we are not looking at the sign, just the magnitude: **10** is larger than **5.**

When adding two integers of the same magnitude but different signs, the answer is always zero.

$$- 5 + 5 = 0$$
$$- 15 + 15 = 0$$
$$6 - 6 = 0$$

Another explanation for adding positive and negative integers.

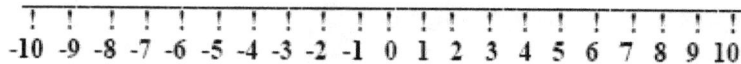

```
-10 -9 -8 -7 -6 -5 -4 -3 -2 -1  0  1  2  3  4  5  6  7  8  9 10
```

Another way of understanding integers is to look at the number line shown above. If you start at **0** and move **3** digits to the left of **0**, i.e. **- 3**, you will be at **- 3**. From **- 3** move another **3** to the left and you will be at **- 6**, i.e **- 3** and **- 3 = - 6**. From **- 6** move another **4** to the left and you will be at **- 10**, *i.e.* **-6** and **– 4 = - 10**.

To the right of **0** all numbers are positive.

+ 4 (**4** digits to the right of **0**) and **+ 3** (**3** digits to the right of **+ 4**) = **+ 7**
+ 4 + 3 = 7

Positive means move to the right; negative means move to the left.

Adding **+7** and **- 3**
From **0** move **7** to the right, i.e. **+ 7**
From **+7** move **3** to the left, i.e. **- 3** and you end up at **4**
+7 - 3 = 4, i.e. same as **+ 4**

Another explanation for adding positive and negative integers.

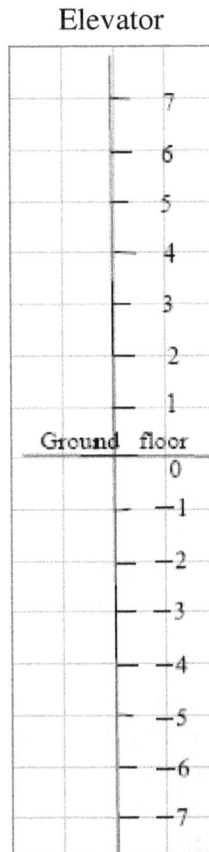

Elevator

```
          7
          6
          5
          4
          3
          2
          1
Ground floor
          0
         −1
         −2
         −3
         −4
         −5
         −6
         −7
```

Imagine an elevator with **7** floors above the ground floor and **7** floors below the ground floor. The floors above the ground floor are positive and the floors below the ground floor are negative If we go *3* floors down, we would be at −**3**. From −**3** if we went another **4** floors down i.e.−**4**, we would be at - **7**

$$-3 \text{ and } -4 = -7 \text{ or}$$
$$-3-4 = -7$$

If we went **4** floors down, i.e.− **4** and from − **4** we went **7** floors up, i.e. + **7**, we would be **3** floors above ground, i.e. +**3**.

$$-4 + 7 = 3$$

When adding a series of positive numbers and negative integers, first add all the positive integers and then add all the negative integers or vice versa and then calculate as previously.

Example

Add **4 - 20 + 30 - 15**

$$4 + 30 \;=\; 34 \qquad \textit{(adding the positive integers)}$$
$$-20 - 15 \;=\; -35 \qquad \textit{(adding the negative integers)}$$
$$34 - 35 \;=\; -1$$

Practice

1. Add
   a) **4 - 5 + 10 - 14**
   b) **- 20 - 14 + 80 - 15**
   c) **20 + 30 - 25 - 5**
   d) **- 3 + 33 - 14 + 50**
   e) **5 - 6 + 35 + 45**
   f) **- 10 + 7 - 7 + 10**
   g) **7 + 8 + 9 - 7 - 8**
   h) **- 25 + 32 - 5 - 8**

Rules for multiplying and dividing integers
Positive multiplied by positive equals positive
Positive multiplied by negative or negative multiplied by positive equals negative
Negative multiplied by negative equals positive

Brackets are used to indicate multiplication

Examples

   a) **5 x 5 = 25** or **(5)(5) = 25** *When there is no sign between the brackets that means multiply.*
   b) **(4)(- 5) = - 20** (*+ multiplied by - equals -*)
   c) **(- 4)(5) = - 20** (*- multiplied by + equals -*)
   d) **(- 5)(-6) = 30** (*- multiplied by - equals +*)
   e) **(- 8)(- 3)(- 4) = - 96** (*- 8x- 3= 24; 24x - 4 = - 96*)
   f) **(7)(-6)(- 3)(- 2) = - 252**

When multiplying an odd number of negative integers, the answer will always be negative.

$$(- 2)(- 2)(- 2) \;=\; -8$$

When multiplying an even number of negative integers, the answer will always be positive.

$$(- 3)(- 2)(- 5)(-2) \;=\; 60$$

Practice 2

Evaluate

    a)   (- **8**)(- **9**)(- **2**)(- **4**)
    b)   (**8**)(- **4**)(- **5**)(**2**)
    c)   (- **7**)(- **4**)(**5**)(**8**)
    d)   (**9**)(**8**)(- **7**)
    e)   (- **6**)(**4**)(- **5**)(- **3**)

<u>The rules for division are the same as for multiplication</u>
Positive divided by positive equals positive
Positive divided by negative or negative divided by positive equals negative
Negative divided by negative equals positive

A division is shown as $\dfrac{12}{7}$ means **12** divided by **7**. (**12** *is the numerator and* **7** *is the*

denominator.)

Examples

    a) $\dfrac{14}{7} = 2$ *(Positive divided by positive equals positive; so* **2** *is positive)*

    b) $\dfrac{14}{-7} = -2$ *(Positive divided by negative equals negative)*

    c) $\dfrac{-14}{7} = -2$ *(Negative divided by positive equals negative)*

    d) $\dfrac{-14}{-7} = 2$ *(Negative divided by negative equals positive)*

Practice 3

Evaluate

    a) $\dfrac{75}{-15}$

    b) $\dfrac{-15}{5}$

    c) $\dfrac{-20}{-5}$

    d) $\dfrac{45}{9}$

    e) $\dfrac{-16}{4}$

    f) $\dfrac{25}{-5}$

CHAPTER 2

FRACTIONS, DECIMALS AND PERCENT

Fractions

A fraction is a part of a whole.

If a pizza was divided into four equal parts and one part was eaten, then a fraction or one quarter of the pizza was eaten. See *Fig 1*. This can be represented in fraction form $\frac{1}{4}$ *(one part out of a total of four parts was eaten; the part eaten is the numerator and the total number of parts is the denominator)*

In this example, **1** is the numerator and **4** is the denominator.

Fig 1

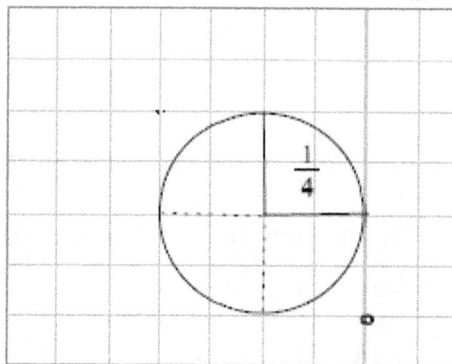

Suppose a pizza was divided into **8** slices and Tom ate **3** slices (*Fig 2*), what fraction of the pizza did Tom eat?

Tom ate **3** out of **8** slices, so Tom ate $\frac{3}{8}$ of the pizza.

(The numerator is the number of slices eaten and the denominator is the total number of slices).

Fig. 2

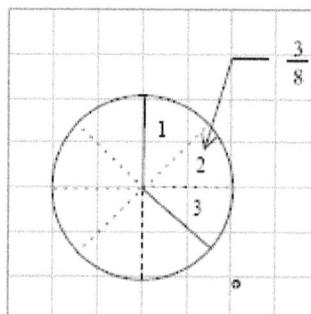

Example

A cake is divided equally amongst **5** people. How much does each person get?

Each person gets $\frac{1}{5}$ of the cake (*1 piece out of **5** pieces*)

If the numerator and denominator of a fraction are multiplied by the same number, the fraction is unaltered.

Example

Multiply the numerator and denominator of $\frac{3}{4}$ by **4**

$$\frac{3 \times 4}{4 \times 4} = \frac{12}{16} : \frac{3}{4} \text{ and } \frac{12}{16}$$ are equivalent fractions (*they are the same.*)

If a cake was divided into **4** pieces and someone was given **2** pieces, and another same size cake was divided into **16** pieces and another person was given **8** pieces, both persons would receive the same amount of cake; the second person would receive more pieces but they would be smaller pieces because the cake was divided into **16** pieces instead of **4**).

The shaded area is half of the square with **16** pieces and the shaded area also corresponds to **8** out of **16**. (See *Fig 3*).

Hence $\frac{1}{2} = \frac{8}{16}$

$\frac{1}{2}$ and $\frac{8}{16}$ are equivalent fractions.

Fig 3

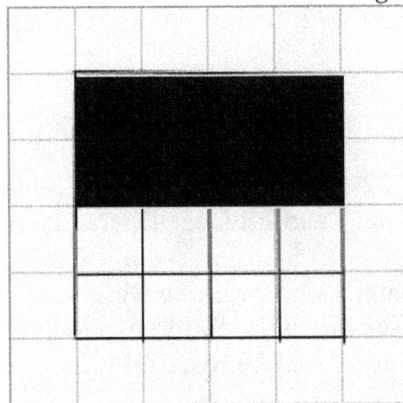

If the numerator and denominator of a fraction are divided by the same number, the fraction remains unaltered.

Example

Divide the numerator and denominator of $\frac{12}{16}$ by **4**

$$\frac{12 \div 4}{16 \div 4} = \frac{3}{4} \quad (\div \text{ is the division sign; } \mathbf{12} \text{ divided by } \mathbf{4} = \mathbf{3} \text{ and } \mathbf{16} \text{ divided by } \mathbf{4} = \mathbf{4})$$

$\frac{12}{16}$ is equal to $\frac{3}{4}$

To express a fraction in its lowest terms, divide the numerator and denominator by the highest factor that is common to both the numerator and denominator.

Example

Reduce $\frac{12}{16}$ to its lowest term; divide numerator and denominator by **4**.

$$\frac{12 \div 4}{16 \div 4} = \frac{3}{4}$$

Example

Reduce to lowest terms $\frac{20}{25}$ divide numerator and denominator by **5** = $\frac{4}{5}$

Example

$\frac{21}{105}$ divide by **3** = $\frac{7}{35}$ divide by **7** = $\frac{1}{5}$ *(The fraction is in its lowest term because no further division is possible)*

or $\frac{21}{105}$ divide by **21** = $\frac{1}{5}$ *(**21** is the highest common factor explained later)*

Practice 1

1. (a) A cake was divided into **12** equal pieces. Mary ate **3** pieces and John ate **4** pieces. What fraction of the cake did Mary and John eat separately and what fraction of the cake did they eat together?
    b) In a bus with a seating capacity of **30**, there were **5** children and the rest were adults. What fraction of the passengers were children, and what fraction were adults?
    c) A flower bed had **5** marigolds and **20** roses. Represent each flower type as a fraction of the total number of flowers.
    d) In a bag there are **3** red marbles, **6** green marbles, **9** yellow marbles, what fraction of each color is in the bag?

2. Reduce to lowest terms
    a) $\frac{24}{32}$ b) $\frac{32}{36}$ c) $\frac{35}{42}$ d) $\frac{25}{30}$ e) $\frac{12}{18}$ f) $\frac{15}{35}$ g) $\frac{18}{34}$ h) $\frac{34}{36}$

Before finding the factors of numbers, knowledge of the divisibility of numbers is required.

Rules for Divisibility of numbers

If a number ends in an even number or **0**, then the number is divisible by **2**.

    22, 34, 46, 58, 50 are all divisible by **2** because they all end in **2, 4, 6, 8** which are even numbers.

If the sum of the digits of a number is divisible by **3**, then the number is divisible by **3**

    549 is divisible by 3 since **5 + 4 + 9 =18** and **18** is divisible by **3**

    243 is divisible by 3 because **2 + 4 + 3 = 9** and **9** is divisible by **3**

If a number ends in **0** or **5**, then the number is divisible by **5**

    20, 25, 30 are all divisible by **5**

If the sum of the digits are divisible by **9**, then the number is divisible by **9**

    6957 is divisible by **9** as the sum of the digits **6+9+5+7= 27** and **27** is divisible by **9**

    If a number is divisible by **9** then the number is also divisible by **3** because **3** goes into **9** or **9** can be divided by **3**

There are other divisibility rules, but the ones above are most commonly used.

Factors

    The factors of **42** are **7, 3** and **2** because **7 x 3 x 2 = 42**

    The factors of **20** are **5, 2** and **2** (**5 x 4 = 20** but **4 = 2 x 2**)

    The factors of **36** are **3, 2, 3** and **2** (**6 x 6 = 36**, but **3 x 2= 6**

Example

Find the factors of **105**

    Since **5+1 = 6** and **3** goes into **6**; **3** is a factor of **105**

    To find the other factor, divide **105** by **3**, which is **35**

    3 x 35 = 105 but **35 = 7 x 5**

    The factors of **105** are **7, 5** and **3**   (*7 x 5 x3 = 105*)

Using a factor tree to find factors

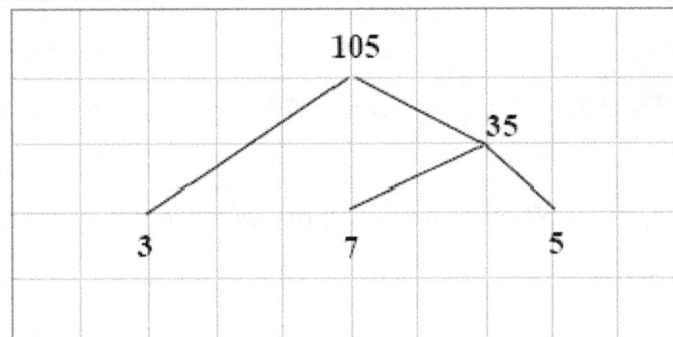

From the factor tree, the factors of **105** are **3** and **35**. The factors of **35** are **7** and **5**, so the factors of **105** are **3, 7** and **5**

Prime and composite numbers

All numbers have a factor of **1**

        23 x 1 = 23
        45 x 1 = 45

Excluding **1** as a factor, if a number has no factors, then it is a prime number.

A composite number is a number with factors.

The following numbers are listed as prime or composite:

1,	2,	3,	4,	5,	6,	7,	8,	9,	10,	11,	12,	13,	14,	15,	16,	17,	18,	19,	21
p	p	p	c	p	c	p	c	c	c	p	c	p	c	c	c	p	c	p	c

Example

Find the factors of **333**

Since **3 + 3 + 3 = 9**, and **3** goes into **9** three times, **3** is a factor of **333**
Divide **333** by **3** to get the other factor (**111** is the other factor)
Let us start with **3** and **111** on the factor tree.

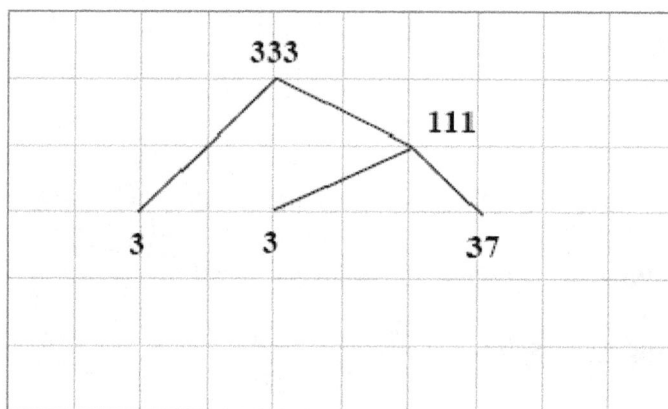

The factors of **111** are **3** and **37** (**37** has no factors; **37** is a prime number). The factors of **333** are **3, 3** and **37**.

Notice that the factors 3, 3 and 37 at the bottom of the factor tree are all prime numbers.

Practice 3

Find the factors to the following, using the factor tree and the divisibility rules.

a) **924** *(find the first factor using the divisibility rule and continue till only prime numbers are left at the bottom)*   b) **987** c) **345** d) **729** e) **1024** f) **1255** g) **782**

Lowest or least common multiple (LCM)

The lowest common multiple of **2, 4** and **12** is **12** because **2** goes into **12** six times; **4** goes into **12** three times and **12** goes into **12** once.

       To calculate the LCM of **2, 4** and **12**, take the multiples of **2, 4** and **12**

The multiples of **2** are **2, 4, 6, 8, 12, 14, 16**	Row **1**
The multiples of **4** are **4, 8, 12, 16**	Row **2**
The multiples of **12** are **12, 24, 36**	Row **3**

12 is the lowest number common to all three rows, so **12** is the LCM of **2, 4** and **12**

The lowest common multiple of **6, 9** and **18** is **18**

       To calculate the LCM of **18**, take the multiples of **6, 9** and **18**

The multiples of **6** are **6, 12, 18, 24**	Row **1**
The multiples of **9** are **9, 18, 27**	Row **2**
The multiples of **18** are **18, 36**	Row **3**

The lowest number that is common to all three rows is **18**, so **18** is the LCM.

Example

Find the LCM of **24** and **36**

       The multiples of **24** are **24, 48, 72, 96**

       The multiples of **36** are **36, 72**

       72 is the LCM

Another method to find the LCM is to find the factors of the numbers.

To find the LCM of **24** and **36**, take the factors of the two numbers.

       The factors of **24** are **2, 2, 3** and **2**

       The factors of **36** are **2, 2, 3** and **3**

The factors that are common to both numbers are **2, 2** and **3**. There is a **2** and **3** that is not common to both numbers.

The LCM is **2 x 2 x 3 x 2 x 3 = 72.** (Multiply the common numbers once and then numbers that are not common).

Example

Find the LCM of **2, 6, 8** and **12**

2 has no factors so we write **2**	Row **1**
The factors of **6** are **2** and **3**	Row **2**

The factors of **8** are **2, 2** and **2**   Row **3**
The factors of **12** are **2, 2** and **3**   Row **4**
2 is common to all **4** rows, so **2** is used once.
2 is common to rows **3** and **4** so another **2** is used once.
3 is common to rows **2** and **4** so **3** is used once.
2 in row three is left, so **2** is used again.

The LCM is **2** x **2** x **3** x **2 = 24**

To solve the same example, using the table below:

No factors of 2	2						
Factors of 6 are	2		3				
Factors of 8 are	2	2		2			
Factors of 12 are	2	2	3				

Use each column once   **2 x   2 x 3 x 2**

Use the factors that are common to the numbers once. Then use the remaining factors that are not common.

Example

Find the LCM of **12, 18, 24**

Factors of 12 are	2	2	3				
Factors of 18 are	2		3	3			
Factors of 24 are	2	2	3		2		

The first **2** is common to all three numbers, so use this **2** once.
The second **2** is common to **12** and **24**, so use this **2** once.
The first **3** is common to **12, 18** and **24**, so use **3** once.
The remaining **3** and **2** are not common, so use them once, each.

The LCM is **2** x **2** x **3** x **3** x **2 = 72**

Example

Find the LCM of **3, 5** and **7**

        Since all three are prime numbers, the LCM is **3 x 5 x 7 = 105** because there are no
        common factors.

Example

Find the LCM of **3, 7** and **15**

        The factors of **15** are **5** and **3**
        No factors for **3**; write **3**
        No factors for **7**; write **7**
        Use **3** only once

The LCM = **3 x 5 x 7 = 105**

Practice 4

Find the LCM using factors.

a) **5, 8, 9**   b) **6, 8, 12**   c) **4, 7, 13** d) **35, 60** e) **7, 45**   f) **5, 6, 45** g) **3, 9, 10** h) **4, 3, 12**

Adding and subtracting fractions

Fractions can be added or subtracted if the denominators are the same.

Example

$$\frac{1}{3} + \frac{1}{3} = \frac{2}{3} \text{ (add the numerators; the denominators remain the same)}$$
$$\frac{1}{5} + \frac{2}{5} = \frac{3}{5}$$

When the denominators are not the same, the fractions can be added by making the denominators the same.

Example

$$\frac{1}{3} + \frac{2}{5} \text{ (The LCM of 3 and 5 is 15, so make both denominators equal to 15. Multiply}$$

        the numerator and denominator of the first fraction by **5** *and the numerator*
        and denominator of the second fraction by **3**)

$$\frac{5 \times 1}{5 \times 3} + \frac{2 \times 3}{5 \times 3} = \frac{5}{15} + \frac{6}{15} = \frac{5+6}{15} = \frac{11}{15}$$

Example

$$\frac{1}{6} + \frac{2}{3} \text{ (The LCM is 6; the first denominator is already 6, so that remains the same)}$$

$$\frac{1}{6} + \frac{2 \times 2}{3 \times 2} = \frac{1+4}{6} = \frac{5}{6}$$

Note

$\frac{-5}{6}$ or $\frac{5}{-6}$ or $-\frac{5}{6}$ are all the same because negative divided by positive, or positive divided by negative equals negative.

Example

$$\frac{-3}{4} + \frac{2}{3} = \frac{3 \times -3}{3 \times 4} + \frac{4 \times 2}{4 \times 3} = \frac{-9}{12} + \frac{8}{12} = \frac{-9+8}{12} = \frac{-1}{12} \text{ or } -\frac{1}{12}$$

Practice 5

Add the following fractions.

a) $\frac{2}{8} + \frac{3}{5}$ b) $\frac{-3}{7} + \frac{2}{9}$ c) $\frac{2}{3} - \frac{3}{4}$ d) $\frac{3}{9} - \frac{1}{3}$ e) $\frac{3}{5} - \frac{1}{3}$ f) $\frac{3}{4} + \frac{2}{3}$ g) $\frac{5}{8} - \frac{3}{4}$

Comparing fractions

Fractions with the same denominator are easy to compare. The fraction with the higher numerator is the greater fraction.

If the denominators are not the same, use the LCM to compare them.

$\frac{3}{5}$ and $\frac{2}{3}$ *The LCM is **15**; multiply the first fraction by **3** and the second fraction by **5** and then compare the numerators.*

Example

Which fraction is greater: $\frac{5}{8}$ **or** $\frac{2}{3}$?

The LCM of **8** and **3** is **24**. Make both denominators **24** by multiplying the first fraction by **3** and the second by **8**.

$$\frac{3 \times 5}{3 \times 8} = \frac{15}{24}$$

$$\frac{2 \times 8}{3 \times 8} = \frac{16}{24} \quad (\frac{16}{24} \text{ is } \frac{2}{3})$$

$\frac{2}{3}$ is larger than $\frac{5}{8}$

Practice 6

Which fraction in each pair of fractions is greater?

a) $\dfrac{2}{3}$ or $\dfrac{3}{4}$ b) $\dfrac{7}{8}$ or $\dfrac{9}{16}$ c) $\dfrac{3}{5}$ or $\dfrac{5}{8}$ d) $\dfrac{3}{8}$ or $\dfrac{3}{5}$

The highest common factor or HCF

The highest or greatest common factor is the highest factor that is common to a group of numbers.

Example

The highest common factor of **4, 6, 12** and **16** is **2**
   The factors of **4** are **2** and **2**
   The factors of **6** are **3** and **2**
   The factors of **12** are **2, 2** and **3**
   The factors of **16** are **2, 2, 2** and **2**

The factor that is common to all the numbers is **2** (only **2** is common to all the numbers).

Example

  Find the HCF of **252** and **90**

Another way of finding factors besides the factor tree is to use repeated divisions.

2	252
2	126
3	63
3	21
	7

When **252** is divided by **2**, the answer is **126** (the second row)
When **126** is divided by **2**, the answer is **63** (third row)
When **63** is divided by **3**, the answer is **21** (fourth row)
When **21** is divided by **3**, the answer is **7** (fifth row)
Always use the lowest factor when dividing. (**252** can be divided by **4**, but **2** was used to start)
The factors of **252** are <u>**2**</u>, **2**, <u>**3**</u>, <u>**3**</u> and **7**

Find the factors of **90** using repeated divisions.

	2	90		
	3	45		
	3	15		
		5		

The factors of **90** are **2**, **3**, **3** and **5**
The factors that are common to **252** and **90** are **2, 3** and **3**
The HCF = **2x3x3 = 18**
To reduce $\dfrac{90}{252}$ to its lowest terms, divide numerator and denominator by **18** = $\dfrac{5}{14}$

Practice 7

Reduce to lowest terms using repeated divisions to find the factors, then find the HCF and use the HCF to reduce to lowest terms.

a) $\dfrac{18}{256}$ b) $\dfrac{25}{40}$ c) $\dfrac{27}{69}$ d) $\dfrac{48}{57}$ e) $\dfrac{92}{146}$

f) $\dfrac{11}{55}$ g) $\dfrac{35}{105}$ h) $\dfrac{39}{87}$ i) $\dfrac{99}{333}$

b) Find the factors in **7a** (numerator and denominator) using the knowledge of divisibility rules and the factor tree, then find the HCF.

Improper fractions

An improper fraction is a fraction where the numerator is greater than the denominator.
$\dfrac{7}{5}$, $\dfrac{9}{8}$, $\dfrac{11}{8}$ are improper fractions.
Fractions should not be left in the improper form but should be converted into a mixed number.

Example

		1	Quotient		
Divisor					
	4	7			
		-4			
Remainder	3		□		

Page 17

To convert $\frac{7}{4}$ into a mixed number, use long division as shown above to divide **7** by **4**, **4** goes once into **7**, leaving a remainder of **3**. **4** is the divisor, **1** is the quotient and **3** is the remainder.

$\frac{7}{4}$ converted into a mixed number is $1\frac{3}{4}$ or **Quotient** $\frac{\textbf{Remainder}}{\textbf{divisor}}$

Convert $\frac{9}{4}$ into a mixed number.

4 goes twice into **9** leaving a remainder of **1** (**2** is the quotient, **4** is the divisor, **1** is the remainder

$\frac{9}{4} = 2\frac{1}{4}$ *(Note the divisor **4** stays in the denominator in both the mixed number and the improper fraction)*

Example

$$\frac{11}{4} = 2\frac{3}{4}$$

Example

$$\frac{34}{6} = 5\frac{4}{6}$$

Example

$$\frac{45}{7} = 6\frac{3}{7}$$

Practice 1

Convert to a mixed number.

a) $\frac{43}{6}$ b) $\frac{37}{5}$ c) $\frac{25}{3}$ d) $\frac{67}{5}$

e) $\frac{27}{7}$ f) $\frac{63}{8}$ g) $\frac{87}{9}$ h) $\frac{51}{7}$

To convert from a mixed number to an improper fraction

$6\frac{3}{7}$, multiply **7** by **6** and add the remainder **3** to get the numerator $7 \times 6 = 42 + 3 = 45$

(**45** is the numerator) **:** the divisor **7** stays the same

$6\frac{3}{7} = \frac{45}{7}$ *(Multiply the quotient by the divisor and add the remainder to get the numerator; the divisor stays the same)*

Example

$$4\frac{3}{4} = \frac{19}{4}$$

Example 7

$$5\frac{3}{8} = \frac{43}{8}$$

Example

$$8\frac{7}{8} = \frac{71}{8}$$

Example

$$9\frac{5}{8} = \frac{77}{8}$$

Practice 2

Convert to an improper fraction

a) $4\frac{3}{4}$ b) $3\frac{5}{8}$ c) $5\frac{7}{8}$ d) $5\frac{6}{7}$ e) $6\frac{3}{4}$ f) $8\frac{5}{8}$

g) $7\frac{7}{8}$ h) $7\frac{2}{3}$ i) $5\frac{3}{7}$ j) $9\frac{4}{5}$ k) $11\frac{3}{4}$

Multiplying and dividing fractions

To multiply fractions, multiply the numerators to get a new numerator and multiply the denominators to get a new denominator.

$$\frac{3}{7} \times \frac{4}{5} = \frac{12}{35}$$ *(The multiplication sign is only shown once, although there are two multiplications)*

Example

Multiply $\frac{7}{5} \times \frac{4}{3} \times \frac{9}{6} = \frac{252}{90} = \frac{14}{5} = 2\frac{4}{5}$ ($\frac{252}{90}$ *reduced to lowest terms* $= \frac{14}{15}$)

To divide fractions, turn the second fraction upside down or interchange the numerators and denominators and then multiply.

Example

$$\frac{2}{3} \div \frac{5}{8}$$

$$= \quad \frac{2}{3} \times \frac{8}{5} = \frac{16}{15} = 1\frac{1}{15}$$

Practice 3

Multiply or divide as the case may be

a) $\frac{2}{4} \times \frac{3}{6}$ b) $\frac{3}{7} \div \frac{4}{5}$ c) $\frac{6}{5} \times \frac{5}{7}$ d) $\frac{7}{11} \div \frac{5}{6}$

e) $\frac{18}{25} \times \frac{5}{7}$ f) $\frac{13}{17} \div \frac{7}{11}$ g) $\frac{6}{7} \div \frac{11}{7}$ h) $\frac{17}{21} \div \frac{15}{21}$

Decimals

When the number "one" is divided into ten equal parts, each one tenth is represented by the decimal **0.1** *(decimal point or point one)*

$$\frac{1}{10} = 0.1 \text{ (decimal point one or decimal one)}; \ 0.1 \text{ is one tenth of the number one}$$

Adding ten tenths would equal one (**0.1 + 0.1 + 0.1 + 0.1 + 0.1 + 0.1 + 0.1 + 0.1 + 0.1 + 0.1 = *1***)
Adding **0.1 + 0.1 = 0.2**
Adding **0.2 + 0.3 = 0.5**

Example 1

```
   Add
   0.3
  +0.5
  +0.7
   1.5
```

When adding decimals, the decimal points have to be aligned vertically and then added the same as any other addition, with carry overs. In example 1, one is carried over to the ones column. The same applies when subtracting decimals. (Example 2: the decimal points are aligned vertically.)

Example 2

```
    4.5
   -2.3
  = 2.2
```

Decimal places

0.23 (**2** is in the tenths place and **3** is in the hundredths place)
0.256 (**6** is in the thousandths place)

When rounding **0.256** to the nearest tenth, look to the digit to the right of the tenths place, which is **5** (hundredths place). If this digit is **5** or more, round up, which means that **0.2** becomes **0.3**. All the digits to the right are discarded. The answer is **0.3**.

Example 3

Round **0.237** to the nearest tenth. Since the hundredths digit (**3**) is **4** or less, the tenths digit (**0.2**) remains the same. The answer is **0.2**. All the other digits to the right are discarded.

Example 4

Round **0.347** to the nearest hundredth. (The hundredths digit is **4**; the digit to the right of **4** is **7** or the thousandths digit. Since this is **5** or higher, round the hundredths digit up to **5**. The answer is **0.35**. The thousandths digit is discarded.

Round **0.678** to the nearest hundredth. The answer is **0.68**.

Round **0.534** to the nearest tenth. The answer is **0.5**.

Round **0.3579** to the nearest thousandth. (answer **0.358**)

Practice

1. Round to the nearest tenth
   a) **0.3578** b) **0.846** c) **0.278** d) **0.4789** e) **0.2784**

2. Round to the nearest hundredth
   a) **0.7452** b) **0.2753** c) **0.5482** d) **0.3856** e) **0.1468**

3. Round to the nearest thousandth
   a) **0.87452** b) **0.8744** c) **0.58939** d) **0.3567** e) **0.5628**

Decimal point

The decimal point is always after the last digit of any whole number although it is never shown.

Example 5
987.
65.
400.

Decimal form

796.2 is written in decimal form and reads seven hundred and ninety six and two tenths. (**796** *is the whole number and* **.2** *is the decimal portion*)

If **796.2** is multiplied by **10**, the decimal point is moved to the right by one place so that **796.2** becomes **7962**.

>   **7.984** x **10** = **79.84**
>   **7.984** x **100** = **798.4**
>   **7.984** x **1000** = **7984**   *(The decimal point moves to the right when multiplying by multiples of* **10**, *by the number of zeros)*

Similarly when a number is divided by multiples of **10**, the decimal point moves to the left by the number of zeros:

>   **7984** ÷**10** = **798.4**
>   **7984** ÷**100** = **79.84**
>   **7984** ÷ **1000** = **7.984**

Multiplying decimal numbers

Example 6

45.2 x **7.2** = **325.44** (**452** x **72** = **32544** without any decimal points, but because there is one decimal point to the left of **2** in each number, the decimal point has been moved to the left by **2** places.

Example 7

4.52 x **7.2** = **32.544** (multiplying without decimals **452** x **72** = **32544** but since there are **2** decimal places in **4.52** starting with **2** and moving to the left to **5** and there is one decimal place in **7.2** from the right of **2** to the left of **2** ; adding **2** + **1** = **3** decimal places; so the decimal place has to be moved **3** places to the left of **32544** as shown in this example).

Example 8

>   **452** x **7.2** = **3254.4**
>   **8.65** x **780** = **6747**   *(Without the decimal the product is* **674700***)*
>   **6.25** x **7.35** = **45.9375**   *(Without the decimal the product is* **459375***)*

Practice 4

Multiply the following as though there were no decimals, and then insert the decimal point as shown above.

a) **2.54 x 6.3**   b) **6.7 x 6.8**   c) **8.9 x 987**   d) **67.5 x 34**   e) **65.2 x 76.9**
f) **300 x 3.5**   g) **410 x 6.25**

<u>Exponents and Scientific notation</u>

$10 \times 10 = 100 = 10^2$ (**2** is the exponent and **10** is the base; **2** represents the number of times **10** is multiplied by itself)
$10 \times 10 \times 10 = 1000 = 10^3$ (**10** is multiplied by itself three times)
$10 \times 10 \times 10 \times 10 = 10000 = 10^4$ (**10** is multiplied by itself four times)

Standard form	Exponential form
10	10^1
100	10^2
1000	10^3

If the exponent is negative as in 10^{-3} ; this means $\dfrac{1}{10^3} = \dfrac{1}{1000} = 0.001$ *(because there are 3 zeros the decimal which was to the right of 1 has been moved 3 places to the left since this is a division)*

$$10^{-1} = \frac{1}{10^1} = \frac{1}{10} = 0.1$$

$$10^{-2} = \frac{1}{10^2} = \frac{1}{100} = 0.01$$

$$10^{-3} = \frac{1}{10^3} = \frac{1}{1000} = 0.001$$

Example 1

$$728 \times 10^2 = 72800$$

Example 2

$$728 \times 10^{-2} = 728 \times \frac{1}{10^2} = \frac{728}{1} \times \frac{1}{100} = \frac{728}{100} = 7.28$$

Points to remember

(Any number written is always in the numerator. **728** means $\dfrac{728}{1}$; *45* means $\dfrac{45}{1}$; etc.)

When a number is multiplied by a multiple of **10**, the decimal point is moved to the right by the number of zeros.

Example 3

7.25 x 10⁴ = 72500 the decimal is moved to the right by four places.

When a number is divided by a multiple of **10**, then the decimal is moved to the left by the number of zeros.

Example 4

76500 ÷ 10⁴ = 7.65 (*The zeros to the right of 5 are discarded as they are after the decimal point*).

When a number is multiplied by **10** raised to a negative exponent, move the decimal to the left by the number of zeros.

Example 5

654 x 10⁻³ = 0.654

Note

Multiplying a number by **0.1** is the same as dividing by **10**
Multiplying a number by **0.01** is the same as dividing by **100**
Multiplying a number by **0.001** is the same as dividing by **1000**

Examples

$$2.34 \times 0.001 = 0.00234$$
$$546.345 \times 0.01 = 5.46345$$
$$76580 \times 0.0001 = 7.658$$

Practice

1. Express the following in decimal form.

    a) 10^{-6} b) 10^{-9} c) 10^{-3} d) 10^{-7}

2. Express in standard form

    a) 10^{5} b) 10^{9} c) 10^{4} d) 10^{7} e) 10^{3}

3. Express the following in exponential form.

    a) **1000**   b) **10000**   c) **10000000**

4. Simplify by expressing in standard or decimal form.

     a) 6.78×10^4 b) 87.3×10^{-2} c) 9780×10^{-3} d) 7.68×10^{-6} e) 2.34×0.0001
     f) 0.007×10^3 g) 0.006×1000 h) 0.078×100 i) 678×0.001 j) 324000×10^{-4}

Scientific notation

Scientific notation is a way of writing large or small numbers in short form. The sun is **93** million miles from the earth. Instead of writing **93000000** which is the standard form, this can be written in scientific notation as follows; 9.3×10^7. The decimal is always placed after the first digit when writing numbers in scientific notation.

A large number such as **868000000000000** can be written as 8.68×10^{14} in scientific notation. A small number such as **0.000000000000258** can be written as 2.58×10^{-13}. (When the decimal in **2.58** is moved thirteen places to the left then the answer is **0.000000000000258**)

Example 1

Express the following in scientific notation.
     a) **4680000** is 4.68×10^6 in scientific notation
     b) **57600000000** is 5.76×10^{10} in scientific notation
     c) **780000000** is 7.8×10^8 in scientific notation
     d) **0.0000000089** is 8.9×10^{-9} in scientific notation
     e) **0.000000000000789** is 7.89×10^{-13} in scientific notation
     f) **0.00876** is 8.76×10^{-3} in scientific notation

Practice 5

Write the following in scientific notation:

     a) **476000000**   b) **5970000**   c) **947000000**   d) **8340000000**
     e) **0.000000078**   f) **0.000000786**   g) **0.0000076**

6. Multiply using the calculator without putting the decimal point and then insert the decimal point

     a) **678.7 x 45**   b) **87.98 x 4.3**   c) **56.89 x 0.89**   d) **987.65 x 93**

PERCENT

A percent is a fraction where the total number of parts is one hundred or the denominator is one hundred. A percent is written by the symbol %.

30% means **30** parts out of a total of **100** and can be written as $\dfrac{30}{100}$

Note that $\dfrac{30}{100}$ is the same as $\dfrac{3}{10}$ and can be written as a decimal **0.3**.

*To convert a fraction to a decimal, first convert the fraction to a percent by multiplying by **100** and then use the result to find the decimal.*

Example 2

Convert $\dfrac{3}{4}$ to a percent and then to a decimal.

$$\dfrac{3}{4} \times \dfrac{100}{1} = \dfrac{300}{4} = 75; \text{ ie } \mathbf{75\%}$$

$$\dfrac{3}{4} = \mathbf{0.75}$$

$$\dfrac{3}{4} = \mathbf{75\%}$$

To convert a percent to a fraction or a decimal, write the meaning of the percent.

Example 3

Convert **45%** into a fraction and a decimal.

$$45\% \text{ means } \dfrac{45}{100} = \mathbf{0.45} \text{ in decimal form.}$$

$$\dfrac{45}{100} = \dfrac{9}{20} \text{ (expressed in a fraction by reducing to lowest terms)}$$

To convert a decimal to a fraction or a percent, write the meaning of the decimal.

$$0.25 \text{ means } \dfrac{25}{100} \text{ which is } \mathbf{25\%}$$

$$\dfrac{25}{100} = \dfrac{1}{4} \text{ reduced to lowest terms which is a fraction.}$$

Practice 7

Fill in the missing values.

	Fraction	Decimal	Percent
1.		**0.55**	
2.	$\frac{2}{5}$		
3.	$\frac{7}{8}$	.	
4.		**0.45**	
5.	$\frac{3}{8}$		
6.		**0.32**	
7.			**46%**

<u>Sale price, purchase price, taxes and tax rate</u>

If an item costs **$100** regularly in a store but during a sale is selling for a discount of **10%** or **10%** off, the item would now sell for **$90.00**. **$90.00** is the discounted price or the <u>sale price.</u> If the sales tax was **15%**, then the price of the item at the check out would be **90.00 + 90 x 0.15 = 90 + 13.5 = 103.50**. **103.50** is the <u>purchase price.</u> **13.50** is the <u>tax</u> on the item. **15%** or **.15** is the <u>tax rate.</u>

<u>Further examples on percent</u>

What is **15%** of **50** ("of" means multiply)?

$$\frac{15}{100} \text{ x } 50 \text{ same as } \frac{15}{100} \text{ x } \frac{50}{1} = 15 \text{ x } \frac{1}{2} = \frac{15}{2} \text{ or } 7\frac{1}{2} \qquad \{ \frac{50}{100} = \frac{1}{2} \}$$

    15% of **50** is **7.5**

If the sales tax is **15%** and an item costs **$50** (in this case there is no discount, so taxes are paid on **$50.00**), what is the purchase price of the item?

    Sales tax = **15%** of **50**; **0.15 x 50 = 7.50**
    The purchase price is the store price, plus **7.50**
    The purchase price = **50 + 7.5 = 57.50**

<u>There is an easier way to find the purchase price.</u>
If an item cost one dollar then the purchase price would be **1.15 (1 + 0.15 = 1.15)**

If the item cost **50** dollars, multiply **50** by **1.15** to get the purchase price.

The purchase price = **50** x **1.15** = **57.5**

Add the tax rate to **1.00** and then multiply by the item cost (**1 + 0.15 = 1.15**)

If the item sells for a discount, subtract the discount rate from **1.00** and then multiply by the item cost.

A coat sells for a discount of **25%**. The store price is **80.00**. What is the price of the coat, with the discount? What is the purchase price of the coat if the sales tax is **15%**?

The price of the discounted coat is (**1 - 0.25**) x **80** = **0.75** x **80** = **60.00**

The purchase price is **60** x **1.15** = **69.00**.

Basic inverse operations

300 x **0.75** = **225** (forward operation ie multiply by 0.75). Start with **300** and after the operation end up with **225**.

$\dfrac{225}{0.75}$ = **300** (reverse operation ie divide by 0.75). Start with **225** and reverse the operation to end up with **300**.

Division is the reverse of multiplication

Example

If an item costing $**200** has been discounted by **15%**, the discounted price is **0.85** x **200** = **170**. If an item after being discounted by **15%** costs $**170**, what is the original price before the discount?

The original price = $\dfrac{170}{0.85}$ = $**200**.

Another example.

A house goes up in value (appreciates) by **15%** a year. If the price today is $**500,000**, what was the price last year? If last year's price is multiplied by **1.15** we get this year's price. Divide this year's price by **1.15** to get last year's price.

Last year's price is $\dfrac{500000}{1.15}$ = $ **434782.61**

This type of problem can also be solved by using ratios which will be dealt with in the next chapter on ratios.

Practice 8

a) A new car is listed at **30,000**. If the sales tax is **12%** and the car is discounted at **15%**, what is the purchase price of the car?

b) Greyhound Coach Lines give a discount of **15%** for Sunday travel. If the sales tax is **15%** and the cost of a regular fare is $**50.00**, what would a ticket cost for Sunday travel?

c) Julie wants to buy shoes. One store is selling a pair of shoes for **$150.00** with a **15%** discount and another store is selling the same shoes at **$145.00** at **10%** off. Which store has a better deal?

d) A real estate agent is paid a commission of **5%** on the sale price of the property. If a house sells for **$500,000** what will the agent's commission be?

e) Jack bought a stereo system at a discount of **15%**. The regular price was **$899**. If the sales tax was **15%**, how much did Jack pay?

f) The bus fare today is **$4.00**. If the price last year went up by **10%**, what was the fare last year?

g) A car depreciates by **18%** a year. If the price today is **$12,500**, what was the price last year?

CHAPTER 3

RATIOS AND RATES

Ratios

When mixing a cake, the ingredients are mixed in a certain ratio. Suppose one cup of raisins is needed for every four cups of flour, then the ratio of raisins to flour is **1:4**.

Let us suppose that a larger cake was being mixed, double the size of the original one, then **2** cups of raisins would be needed for **8** cups of flour. The ratio of raisins to flour would be **2:8**. The ratio is the same in both cases. These two ratios can be written as fractions $\frac{1}{4}$ and $\frac{2}{8}$.

Since the two ratios are the same $\frac{1}{4} = \frac{2}{8}$, these are equivalent fractions as discussed before.

Examples

Find the missing number.

a) $\quad \frac{3}{4} = \frac{\Diamond}{8}$ The missing number is $\frac{8}{4} \times 3 = 6$

Since the missing number is on the right of the equal sign and **8** is also on the right of the equal sign, **8** is the numerator and **4** is the denominator. The result is multiplied by **3**.

b) $\quad \frac{7}{14} = \frac{5}{\Diamond}$ Since the missing number is to the right of the equal sign and **5** is to the right of the equal sign, **5** is the numerator and **7** is the denominator. The result is multiplied by **14**.

$$\Diamond = \frac{5}{7} \times 14$$
$$\Diamond = 10$$

c) $\quad \frac{6}{\Diamond} = \frac{5}{8}$

$\Diamond = \frac{6}{5} \times 8 = 9.6$ Since the missing number is to the left of the equal sign, and **6** is to the left of the equal sign , **6** is the numerator and **5** is the denominator and the result is multiplied by **8**.

Practice 1.

Find the value of $\Diamond$ in each of the following.

a) $\frac{3}{5} = \frac{6}{\Diamond}$ b) $\frac{5}{6} = \frac{7}{\Diamond}$ c) $\frac{6}{8} = \frac{\Diamond}{12}$ d) $\frac{2}{\Diamond} = \frac{4}{7}$

e) $\frac{3}{2} = \frac{\Diamond}{8}$ f) $\frac{8}{9} = \frac{3}{\Diamond}$ g) $\frac{3}{8} = \frac{5}{\Diamond}$

Ratios can be used to solve a number of different problems

If the sales tax is **15%** and an item that is marked at $85 in the store, the purchase price (the price that the customer pays at the check out) will be **$85 x 1.15 = 97.75**

What will the purchase price be on an item that cost **$125** at the store?

With a **15%** sales tax, if an item costs **$100**, then the purchase price would be **$115**.

Write this as a ratio:

$$\frac{\text{item cost}}{\text{purchase price}} = \frac{\text{item cost}}{\text{purchase price}}$$

$$\frac{100}{115} = \frac{125}{\text{purchase price}}$$

$$\frac{125}{100}\text{x}115 = \text{purchase price.}$$

$$\text{purchase price} = \$143.75$$

The ratio method was used to show the different ways of solving the problem.

The easier method is to multiply the item cost by **1.15** to get the purchase price.

$$\text{Purchase price} = \textbf{125 x 1.15} = \$143.75$$

With the same example, if the purchase price is **$143.75**, what is the item cost before sales tax?

$$\frac{\text{item cost}}{\text{purchase price}} = \frac{\text{item cost}}{\text{purchase price}}$$

$$\frac{100}{115} = \frac{\text{item cost}}{143.75}$$

$$\text{item cost} = \frac{143.75}{115} \text{ x } 100$$

$$\text{item cost} = \ \$125$$

Three quantities are given, so the fourth can be calculated.

Or the easier method to find the item cost is to divide the purchase price by **1.15**:

$$\text{Item cost} = \frac{143.75}{115} = \$125$$

Ratios can be used to solve rate and percent problems

Example

45% of what is **60**? Of what means the total (whole) and is means the part.

$$\frac{\textbf{part}}{\textbf{whole}} = \frac{\textbf{part}}{\textbf{whole}}$$

$$\frac{45}{100} = \frac{60}{\textbf{of what}}$$

Of what (total) $= \dfrac{60}{45} \times \textbf{100} = 133.33$

Checking
45% of **133.33**
 0.45 x **133.33** $=$ **59.99**
45% of **133.33** is **59.99** or **60** (checks out)

Example

A car travels **500** km in **8** hours. What distance will the car travel in **6** hours?

$$\frac{500}{8} = \frac{\Diamond}{6} \qquad \Diamond = \frac{6}{8}\textbf{x500} = \textbf{375} \text{ km}$$

$$(\frac{\textbf{km}}{\textbf{h}} = \frac{\textbf{km}}{\textbf{h}})$$

Example

A car travels **400** km in **5** hours. How long will it take for the car to travel a distance of **300** km?

$$\frac{400}{5} = \frac{300}{\Diamond}$$

$$\Diamond = \frac{300}{400} \times \textbf{5} = \textbf{3.75} \text{ hours}$$

Example

Express $\dfrac{3}{4}$ as a percent:

$$\frac{\textbf{part}}{\textbf{whole}} = \frac{\textbf{part}}{\textbf{whole}}$$

$$\frac{3}{4} = \frac{\Diamond}{100} \qquad \text{Line 1}$$

$$\Diamond = \frac{100}{4} \times 3 = 75\%$$

Checking :   From line 1, $\frac{3}{4} = \frac{75}{100}$ or **75%**   (replacing $\Diamond$ with **75**)

Ratios used to solve percentage increase problems

Example

A blade of grass grows from **3** cm to **4** cm in a year. What is the percentage increase?

$$\frac{\textbf{increase in length}}{\textbf{original length}} = \frac{\textbf{increase in length}}{\textbf{100}}$$

The blade increased in length by **4 – 3 = 1** cm.

$$\frac{1}{3} = \frac{\textbf{Increase in length}}{\textbf{100}}$$

The increase in length $= \frac{100}{3} \times 1 = 33.3\%$
or convert $\frac{1}{3}$ to a percent, multiply by **100**

$$\frac{1}{3} \times \textbf{100} = \textbf{33.3\%}$$

The ratio method of solving problems is shown to make you better understand the problem. Solve any problem by whichever method is the easiest.

In the above example, converting the fraction to a percent is the easier method.

More examples on ratios

If the ratio of boys to girls in a classroom is **12: 8**, what percentage of the classroom are boys? What percentage are girls?

If there are **12** boys and **8** girls then there would be **20** students, out of which **12** are boys.
12 out of **20** are boys. This can be written as a fraction $\frac{12}{20}$; multiply by **100** to get the percentage.

$$\text{The percentage of boys } = \frac{12}{20} \times \textbf{100} = \textbf{60\%}$$
$$\text{The fraction of girls} = \frac{8}{20}$$
$$\text{The percentage of girls } = \frac{8}{20} \times \textbf{100} = \textbf{40\%}$$
$$\text{Checks out } \textbf{60} + \textbf{40} = \textbf{100}$$

Calculating the percentage of girls using ratios.

$$\frac{8}{20} = \frac{\textbf{percentage of girls}}{100}$$

Percentage of girls = $\frac{100}{20}x$ 8 = 40%

Example

If one out of **8** provinces speak French and the rest are English speaking, what percentage of the provinces speak English ?

English speaking provinces = $\frac{7}{8}$ x100 = 87.5%

Example

Anita and Brian won a lottery worth **$100,000**. Each person's winning is proportional to the dollar value of their tickets. If Anita spent **$15** and Brian spent **$10**, how much did each receive?

Expressing each person's share of the winnings as a fraction of the total spent on the tickets. The total spent on the two tickets is **15 + 10 = 25**.   Expressing each person's share as a fraction.

Brian's share = $\frac{10}{25}$ x **100000 = 40,000**

Anita's share = $\frac{15}{25}$ x **100000 = 60,000**

Three term ratios

Example

A piggy bank has nickel, dimes and quarters in the ratio of **2:3:4**. If there are **6** nickels, what is the number of dimes and quarters?

Nickels: dimes = **2:3** or $\frac{N}{D} = \frac{2}{3}$ *(N = nickels; D = dimes; Q = quarters)*

$\frac{6}{D} = \frac{2}{3}$ D = $\frac{6}{2}$x 3 = **9** ;

D = **9**

$\frac{D}{Q} = \frac{3}{4}$

$\frac{9}{Q} = \frac{3}{4}$

Q = $\frac{9}{3}$ x 4 = **12**

There are **9** nickels and **12** quarters.

Since we know the ratio of nickels to dimes and the number of nickels are given, we start with that ratio first.

Practice 2

a) Nick's ice cream has three flavors; chocolate, strawberry and vanilla in the ratio of **2:3:4**; if there are **9** ounces of ice cream in a cone, how many ounces of each flavor ?

b) A bag of fruit has oranges apples and bananas in the ratio of **3:7:8**. If there are **24** bananas, what is the number of apples and oranges?

c) In a floral arrangement there are tulips, roses and marigolds in the ratio of **2:5:8**. If there are **24** marigolds, what is the total number of flowers and how many of each type?

d) A cereal contains oats, nuts and bran in the ratio of **2:3:4**. How much of each ingredient is required for the following mixes?
   1) **72**g   2) **180**g   3) **135**g   4) **162**g

e) Rangers in a park tagged and released **120** foxes. Six months later when they captured **20** foxes, **4** had tags. Estimate the fox population.

f) In a bag containing marbles, the ratio of red, green and yellow marbles is **3:4:5** . If there are **120** marbles, how many marbles are there of each color?

g) In Epping forest the ratio of rabbits to foxes is **8:3**. If the total number of foxes and rabbits are **440**, what is the fox population?

3.
a) **30%** of what is **50**?

b) **16%** of **40** = ◊

c) $\dfrac{15}{20} = ◊\%$

d) **45%** of **40** = ◊

e) **75%** of the students in a student council voted. If **450** students voted, how many students were in the council?

4. Sam bought a shirt with a regular price of $**31.00** . The shirt was on sale for a **15%** discount. The taxes were **15%**. Determine each amount.

a) the purchase price.   b) the discount.
c) the taxes.   d) the sale price.

Rates

A rate is a comparison between two quantities, the denominator is one.

• kilometers per hour (kilometers for one hour) ,
• words per minute,
• dollars per kilogram.

The unit rate is obtained by dividing one quantity by the other.

In the first example, kilometers are divided by hours. A car travels **300** kilometers in **4** hours, what is the speed of the car?

The car's speed $= \dfrac{300}{4} = $ **75** kilometers per hour or **75** km/h

($\dfrac{75}{1}$: the denominator is **1** in final answer)

The quantity after the / is the denominator or the quantity after the word per goes in the denominator.

By expressing the rate in English, it is possible to know which quantity is the numerator and which is the denominator.

Examples

- A room in a hotel costs dollars per day, not days per dollar, so you know that dollars is in the numerator and days is the denominator or divisor.
- A typist types *x* number of words per minute, not minutes per word.
- A bag costs *x* number of dollars per bag, not bags per dollar.

Practice

1. Express the following in unit rates.

   a) A typist types **500** words in **5** minutes.
   b) A runner ran **400**m in **55** seconds.
   c) **$3.00** for **4** m of typewritter ribbon.
   d) Andrew worked for **6** hours and earned **$80** while Sara earned **$120** for **11** hours . Who earned a greater hourly rate?
   e) **$6.00** for a **450** g box of cereal.
   f) **$5** for **7** litres of gas.

Once the unit rate is determined, then it is easy to calculate the numerator.

If a car travels **60** km/h, how far will it travel in **6** hours. Multliply the unit rate by the number of hours to get the kilometers travelled.

2.
a) A person types **600** words in **7** minutes; how many words can the person type in **10** minutes?
b) Jack earns **$700** in **8** hours; how much will he earn for a **40** hour week?
c) Sarah filled her gas tank with **45** litres of gas; how far can she drive if her car does **40**km per litre?
d) Seven bus tickets cost **$10.50** . What is the cost of **20** tickets?

3. Hilton hotel charges **$350** for **5** days where as the queens hotel charges **$600** for **8** days. Which is the better rate?

4. Brand A detergent costs **$12.00** for **550** grams and Brand B costs **$7.00** for **300** grams. Which is cheaper?

5. Malik drives at **70** km/h. How far can he drive in **5** hours?

6. An aircraft flies **5000** km in **6** hours, How far will it fly in **5** hours?

CHAPTER 4

MEASUREMENTS

Metric measurements

10 millimeters = **1** centimeter
10 centimeters = **1** decimeter
10 decimeters = **1** meter
10 meters = **1** decameter
10 decameters = **1** hectometer
10 hectometers = **1** kilometer

Of these units, millimeters (mm), centimeters (cm), meters (m) and kilometers (km) are usually used for measurements.

When multiplying numbers with multiples of **10**, add the number of zeros in each number to get the answer.

$$100 \times 100 = 10000 \qquad 10^2 \times 10^2 = 10^4$$
$$100 \times 1000 = 100000 \qquad 10^2 \times 10^3 = 10^5$$
$$1000 \times 1000 = 1000000 \qquad 10^3 \times 10^3 = 10^6$$

When dividing numbers in exponential form with a base of **10**, subtract the exponents in the denominator from the exponent in the numerator.

$$\frac{10^4}{10^2} = 10^2$$

Another way of evaluating the same example is to express it standard form.

$$\frac{10000}{100} = \frac{100}{1} = 100 \quad (\textit{dividing numerator and denominator by } \textbf{100})$$

Example

$$\frac{100000}{100} = 1000 \quad \textit{The } \textbf{2} \textit{ zeros in the denominator cross out } \textbf{2} \textit{ zeros in the}$$

*numerator leaving **3** zeros in the numerator.*

The chart below is useful to convert from one metric unit to another (Linear metric conversion)

km	hm	dm	m	dm	cm	mm
◊	◊	◊	◊	◊	◊	◊
1	**2**	**3**	**4**	**5**	**6**	**7**

Each square represents a multiple of **10**.
1 cm = **10** mm
1 decimeter (dm) = **10** cm
1 m = **100** cm and so on.

Starting with square **1** which is km and moving to the right **2**, **3** and **4** we reach square **4** which is meters. Counting **2**, **3** and **4** (**10x10x10**) =**1000**. That means multiply km by **1000** to get meters (or **1000** meters = **1** km)

From meters to cm and not counting meters are two squares **5** and **6**; so multiply meters by **100** to get cm (or **100** cm = **1** m)

From cm to mm is one square, **7** (not counting **6**), so multiply cm by **10** to get mm (or **10** mm = **1** cm)

To find out the number of mm in one km, start with **7** (do not count **7**) and move to the left to the km square (**1**). There are six squares so there **1000000** mm in **1** km.

Once this is understood the squares would not be necessary.

Example

Convert **5**km to m.
Since there are **1000** meters in **1** km, <u>multiply</u> 5 by **1000** =**5000** m
5 km = **5000** m
To convert **5000** meters to km., <u>divide</u> **5000** by **1000** = 5 km
5000 m = **5** km

<u>When converting from a larger unit to a smaller unit, multiply</u>
<u>When converting from a smaller unit to a larger unit, divide</u>

Example

Convert **5000** mm to meters.
The first step is to find out how many millimeters there are in a meter.
From the chart there are **1000** mm in **1** m. Because we are converting from a smaller unit to a larger unit, divide **5000** by **1000** = **5** meters.

Example

Convert **5.23** km to centimeters.
There are **100000** or 10^5 cm in one km; since we are converting from a larger to a smaller unit, multiply **5.23** by 10^5 = **523000**.

Example

Convert **5567** mm to km
There are **1000000** mm in 1km; since we are converting from a smaller to a larger unit, divide **5567** by 10^6 = **0.005567** km

Note : Dividing any number by 10^6 is the same as multiplying the same number by 10^{-6}

Practice

1. Convert each measurement to meters.
a) **5.89** km   b) **56** mm   c) **473** cm   d) **0.0675** km   e) **8954** mm   f) **7593** cm

2. Convert each measurement to centimeters.
a) **35** m   b) **570** mm   c) **4.75** km   d) **2.63** mm   e) **38.5** m   f) **8.34** mm

3. Fill in the blanks
a) **35** cm = ◊ m   b) **3.75** km = ◊ m   c) **3.45** mm = ◊ km   d) **0.64** cm = ◊ mm
e) **14.7** km = ◊ cm   f) **6.34** km = ◊ m

Area metric conversion

In a rectangle all the angles are **90** degrees.

7 cm

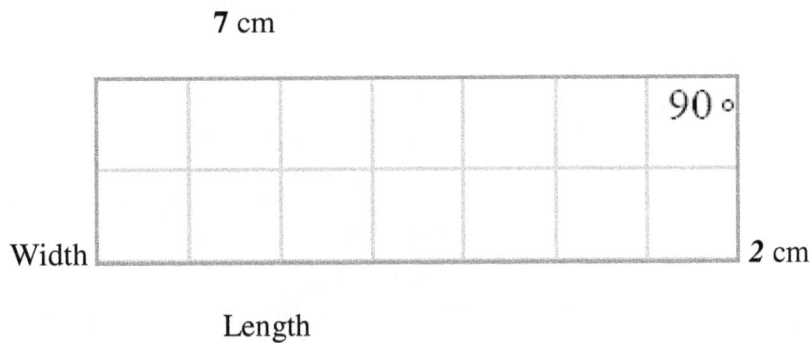

Width   **2** cm

Length

The area of the rectangle is **7x2** cm^2 = **14** cm^2
The area of a rectangle = L x W (Length x Width)

The area of a square = side multiplied by side since both sides are equal

6 cm

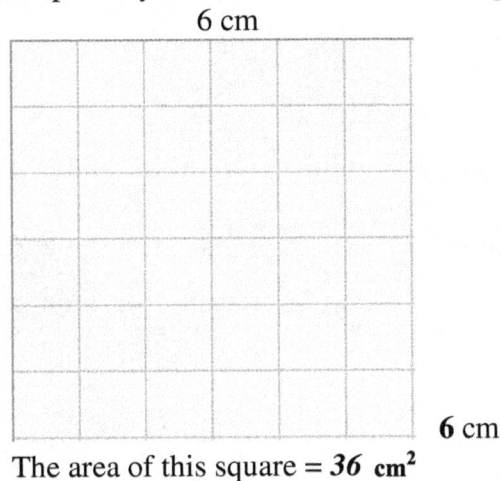

6 cm

The area of this square = **36** cm^2

Converting square cm to square mm.

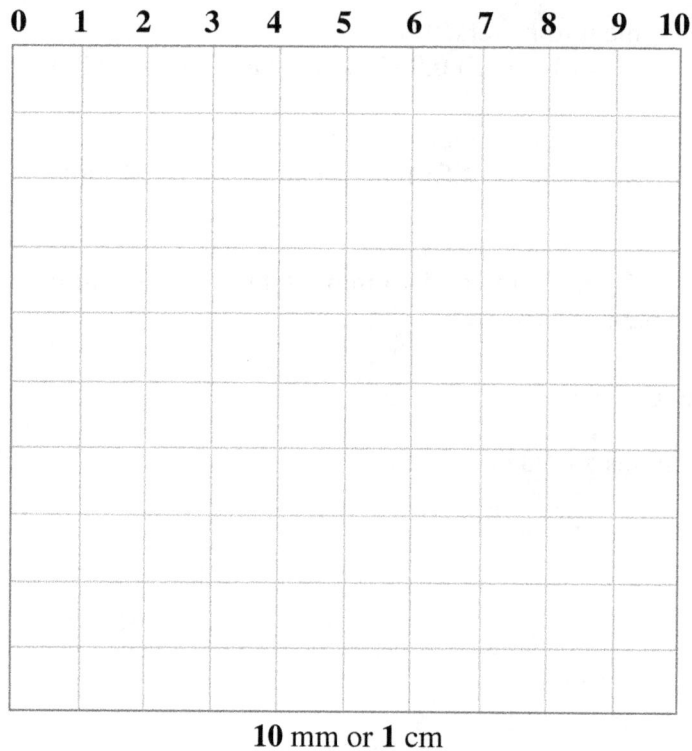

0   1   2   3   4   5   6   7   8   9   10

10 mm or **1** cm

The side of this square is **1**cm or **10** mm.
The area of this square is **1** cm^2.
The area in mm is **10x10 = 100** mm^2.
1 cm^2 = **100** mm^2 .

Example

       1 m =**100** cm
       1 m^2= **100 x 100** cm^2.

Example

       1 m = **1000** mm.
       1 m^2 = **1000 x 1000** mm^2
       1 m^2 = **10^6** mm^2

Example

       1km = **1000** m

$$1 \text{ km}^2 = 1000 \times 1000 \text{ m}^2$$
$$1 \text{ km}^2 = 10^6 \text{m}^2$$

Example

Convert **30** m² to cm²
$$1 \text{ m}^2 = 10^4 \text{ cm}^2$$
$$30 \text{ m}^2 = 30 \times 10^4 \text{ cm}^2$$

Example

Convert **4000** cm² to m²
$$= 4000 \times 10^{-4} \text{ m}^2 \qquad$$ *(Converting from a smaller unit to a larger, divide by 10^4. Refer to the topic on exponents and scientific notation)*
$$= 0.4 \text{ m}^2$$

Example

Convert **5.56** m² to km²
$$= 5.56 \times 10^{-6} \text{ km}^2 \qquad$$ *(Converting from a smaller to a larger unit, divide by 10^6;*
***1000m = 1km ; 1000 x1000 m² = 1 km²**)*

Practice

1. Convert each area to mm²
   a) **4.4** km² b) **6.43** m² c) **54** cm²

2. Convert each area to m²
   a) **589** mm² b) **550** cm² c) **54.4** km²

3. Convert to cm²
   a) **50** mm² b) **45** km² c) **85** m²

Volume metric conversion

The volume of a rectangular prism is equal to length x width x height.
A rectangular prism is a three dimensional rectangular object where the opposite faces are parallel.
Parallel lines or faces do not meet.

A _____ B   AB and CD are parallel lines.

C _____ D

In Fig 1 ABCD and EFGH are parallel faces.

BCGH and ADHE are another pair of parallel faces and finally CDHG and BAEF are also parallel.

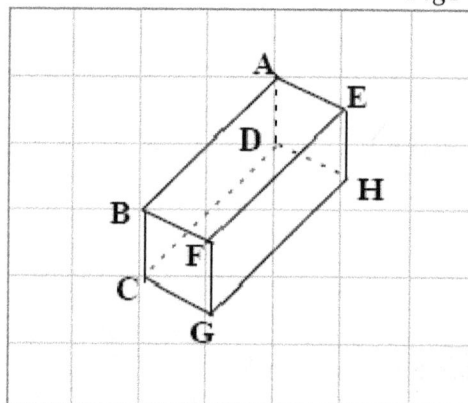

Fig1

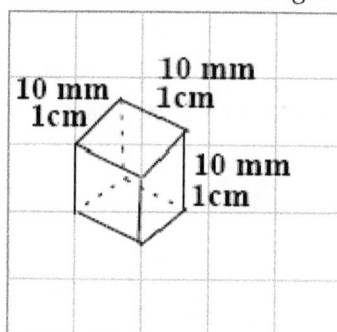

Fig 2

A cubic centimeter is shown in Fig 2. A cubic centimeter is a cube with each side equal to **1 cm** and is written as cm^3.

$$1\ cm^3\ =\ 10x10x10\ mm^3$$

Example

$$1\ km\ =\ 1000\ m$$
$$1\ km^3\ =\ 1000x1000x1000\ m^3\ or\ 10^9\ m^3$$

Example

$$1\ m\ =\ 100\ cm$$
$$1\ m^3\ =\ 100x100x100\ or\ 10^6\ cm^3$$

Example

$$100 \text{ cm} = 1 \text{ m}$$
$$100 \times 100 \times 100 \text{ cm}^3 = 1 \text{ m}^3$$
$$10^6 \text{ cm}^3 = 1 \text{ m}^3$$

Convert **1000** cm^3 to m^3

Since we are converting from a smaller to a larger unit, divide **1000** by **10^6**

$$\frac{10^3}{10^6} = \frac{1}{10^3} = 0.001 \text{ m}^3$$

$$1000 \text{ cm}^3 = 0.001 \text{ m}^3 \quad \text{(Refer to the topic on exponents and scientific notation page 23)}$$

Example

Convert **5000** mm^3 to cm^3

$$10 \text{ mm} = 1 \text{ cm}$$
$$10 \times 10 \times 10 \text{ mm}^3 = 1000 \text{ mm}^3$$
$$1000 \text{ mm}^3 = 1 \text{ cm}^3$$

Since we are converting from a smaller unit to a larger one, divide **5000** by **1000 = 5** cm^3
5000 mm^3 = **5** cm^3

A litre is equal to **1000** cc (cc is the same as cm^3 or cubic centimeter)
A litre is also equal to **1000** ml (millilitres)
1 cc **= 1** ml

Example

How many litres are there in **1** m^3

$$1 \text{ m} = 100 \text{ cm}$$
$$1 \text{ m}^3 = 100 \times 100 \times 100 \ cm^3 = 10^6 \text{ cm}^3$$
$$\mathbf{1 \ m^3 = 10^6 \ cm^3}$$

Since **1** litre is equal to **1000** cc

$$\mathbf{1 \ m^3} = \frac{10^6}{10^3} = 10^3 \text{ litres.}$$
$$1\text{m}^3 = \mathbf{1000} \text{ litres.}$$

Example

Convert **1000** mm^3 to litres.
The first step is to convert mm^3 into cm^3
10 × 10 × 10 mm^3 or **1000**mm^3= **1** cm^3 (but **1**cm^3 = **1** ml)

$1000 \text{ mm}^3 = 1$ ml or 0.001 litre
$1000 \text{ mm}^3 = 0.001$ litre.

Practice 1

a) The height of water in a swimming pool is **2** meters. The length and width of the pool are **8** meters and **6** meters respectively. Calculate the litres of water required to fill the pool. Use scientific notation if necessary.

b) Convert **5000** cm^3 into km^3

c) Convert **5 x 10^6** mm^3 into km^3. Use scientific notation.

d) Convert **2** km^3 into mm^3 Use Scientific notation.

Squares and square roots

$4^2 = 16$ is read as four squared (**4 x 4**)

The square root of **16** is **4**. (What number when multiplied by itself equals **16**; the answer is **4**, so **4** is the square root or the root of **16**)

Squaring a number and then taking the square root of the result leaves the number unchanged. $4^2 = 16'$ $\sqrt{16} = 4$ ($\sqrt{}$ means the square root $\sqrt{16}$ means the square root of **16**; $\sqrt{}$ is the radical sign . The value under the radical sign is called the radicand. In this case **16** is the radicand.

Squaring a number and then taking the square root leaves the number unchanged as these operations are opposite.
$\sqrt{5^2} = \sqrt{25} = 5$ (Squaring **5** and then taking the square root of the result leaves **5** unchanged)
$\sqrt{6^2} = \sqrt{36} = 6$

This will be covered in detail in a later chapter.

Practice

2. Find the square roots of

1. **36** b) **64** c) **81** d) **100** e) **25** f) **9** g) **16**

3. Evaluate
a) $\sqrt{2^2}$ b) $\sqrt{4^2}$ c) $\sqrt{5^2}$ d) $\sqrt{7^2}$ e) $\sqrt{9^2}$

CHAPTER 5

EQUATIONS 1

In an equation the left hand side is equal to the right hand side.

5 = 5 is an equation because the left hand side of the equal sign equals the right hand side of the equal sign.

Whatever is to the left of the equal sign is called the left hand side (LHS) and what is to the right of the equal sign is called the right hand side (RHS). For an equation to be true, both sides must have the same value.

$$4 + 3 = 7 \quad \text{is an equation because the LHS = RHS}$$
$$4 = 7 - 3 \quad \text{The 3 which was on the LHS has been moved to the RHS and is now negative 3 and the equation is still true.}$$
$$4 = 4$$

<u>When a number is moved from the LHS to the RHS or from the RHS to the LHS of an equation, the sign changes</u>

Example

$$x + 3 = 7 \quad \textit{(x is a variable; its value changes according to the equation. To find the value of x, move the 3 to the RHS of the equation)}$$
$$x = 7 - 3$$
$$x = 4$$
$$\text{Checks out } 4 + 3 = 7$$

Example
$$x - 4 = 12$$
$$x = 12 + 4$$
$$x = 16$$
$$\text{Checks out } 16 - 4 = 12$$
$$12 = 12$$

<u>Brackets with no sign in between them means multiply</u>
4(3) means **4** multiplied by **3**
(4)(3) also means **4** multiplied by **3**

Practice

1. Solve for the variable.
a) $x - 7 = 25$ b) $x - 9 = 3$ c) $y + 8 = 3$ d) $z + 5 = 54$ e) $t - 45 = 4$ f) $c + 7 = 56$
g) $h - 56 = 45$ h) $k + 6 = 7$

When x is multiplied by a fraction such as $\frac{4}{3}x$; same as $\frac{4x}{3}$, **4** is the multiplier of x (numerator) and **3** is the divisor of x (denominator)

In equations, the rules for multiplication and division are different from those of addition and subtraction.

Example 1

$$4x \;=\; 12$$

(*4 is the multiplier of x in the LHS of the equation, but becomes the divisor (denominator) when moved to the RHS ; 4x means 4 multiplied by x)*

$$x \;=\; \frac{12}{4}$$

4 is moved to the RHS to find the value of x or to isolate x.

<u>When a number or variable is written, the number or variable is always in the numerator</u>

4 is the same as $\frac{4}{1}$ and x is the same as $\frac{x}{1}$ The numerator is always the multiplier and the denominator is the divisor.

Example 2

$$\frac{x}{4} \;=\; 12$$

4 is the divisor (denominator) on the LHS of the equation, but becomes the multiplier (numerator) when moved to the RHS.

$$x \;=\; 12(4)$$
$$x \;=\; 48$$

<u>The general rule is that the multiplier becomes the divisor and the divisor becomes the multiplier, when moved to the other side of the equal sign</u>

Example 3

$$4x + 3 \;=\; 15 \quad \text{(First move the 3 to the RHS)}$$
$$4x \;=\; 15 - 3$$
$$4x \;=\; 12$$
$$x \;=\; \frac{12}{4}$$
$$x \;=\; 3$$

Example 4

$$\frac{3x}{4} + 4 = 16$$

$$\frac{3x}{4} = 16 - 4$$

$$\frac{3x}{4} = 12$$

$$x = \frac{12(4)}{3}$$

$$x = 16$$

Example 5

$$\frac{3x}{5} = \frac{15}{2}$$

$$x = \left(\frac{15}{2}\right)\left(\frac{5}{3}\right)$$ *(5 is multiplied by whatever is in the numerator and 3 is multiplied by whatever is in the denominator)*

$$x = \frac{25}{2} \quad \text{or } 12.5$$

When there are fractions with multiplications, always reduce to lowest terms first.
$$\left(\frac{16}{4}\right)\left(\frac{15}{3}\right) = 4 \times 5 = 20 \quad (\frac{16}{4} = 4 ; \frac{15}{3} = 5)$$

In equations, whatever operation is done to the left hand side must also be done to the RHS)

Example 6

$$\frac{5x}{3} = \frac{2}{3}$$ *(since the denominator is 3 on both sides, multiply both sides by 3 to eliminate the denominator)*

$$\frac{3(5x)}{3} = \frac{3(2)}{3} \quad (\frac{3}{3} = 1)$$

$$5x = 2$$

$$x = \frac{2}{5}$$

Example 7

$$\frac{4x}{-3} + 7 = 11$$

$$\frac{4x}{-3} = 11 - 7$$

$$\frac{4x}{-3} = 4$$

$$4x = (-3)(4)$$

(note when the -3 is moved to the RHS of the equation the sign does not change because -3 is multiplied by whatever is in the RHS)

$$4x = -12$$
$$x = \frac{-12}{4}$$
$$x = -3$$

Example 8

$$-\frac{4x}{3} = 12$$
$$x = 12\left(\frac{-3}{4}\right)$$
$$x = -9$$

Example 9

$$-x = 4$$

(If all the signs are changed in an equation the equation remains unaltered. <u>Since we do not want a minus in front of the x</u>, change all the signs on both sides of the equation)
This is equivalent to multiplying both sides by - 1

$$x = -4$$
$$(-1)(-x) = (-1)(4)$$
$$x = -4$$

Example 10

The relationship between centigrade and fahrenheit is given by the formula.

$$\frac{c}{5} = \frac{f-32}{9} \quad \textit{Express c in terms of f.}$$

$$c = \frac{5(f-32)}{9} \qquad \frac{f-32}{9} \textit{ is the same as } \frac{(f-32)}{9}$$

Example 11

$$\frac{c}{5} = \frac{f-32}{9} \quad \textit{Express f in terms of c.}$$

$$\frac{9c}{5} = f-32 \quad \textit{(9 was the divisor in the RHS, becomes the multiplier in the LHS)}$$

$$\text{or } f-32 = \frac{9c}{5}$$

$$f = \frac{9c}{5} + 32$$

Practice 2

a) $c = 6r$ express **r** in terms of **c**.

b) $v = 6rh$; (6rh means **6** multiplied by **r** multiplied by **h**) Express **r** in terms of **v** and **h**; express **h** in terms of **v** and **r**.

c) $a = wl$ Express w in terms of **a** and **l**. Express **l** in terms of **a** and **w**.

d) $v = lwh$. Express l in terms of **v**, **w** and **h** ; express **w** in terms of **v**, **l** and **h**; express **h** in terms of **v**, **l** and **w**.

e) $a = \dfrac{bh}{2}$ Express **b** in terms of **a** and **h** ; express **h** in terms of **a** and **b**.

f) Express **y** in terms of **x** and then **x** in terms of **y**.

$\dfrac{5x}{2} + \dfrac{3y}{5} = 10$ (hint ; multiply by LCM)

g) The area of a trapezoid $= \dfrac{h(a + b)}{2}$; $A = \dfrac{h(\textbf{sum of parallel sides})}{2}$

Express the sum of the parallel sides in terms of A and h.

3. Solve for **x**

a) $4x + 5 = 25$

b) $3x - 3 = 12$

c) $5x - 7 = 13$

d) $\dfrac{x}{3} = 12$

e) $\dfrac{x}{4} + 7 = 2$

f) $\dfrac{x}{7} + 10 = 7$

g) $\dfrac{-x}{5} + 7 = 12$

h) $\dfrac{-5x}{3} - 8 = 12$

i) $\dfrac{-3x}{4} - 12 = -6$

Plotting points on a Cartesian coordinate system

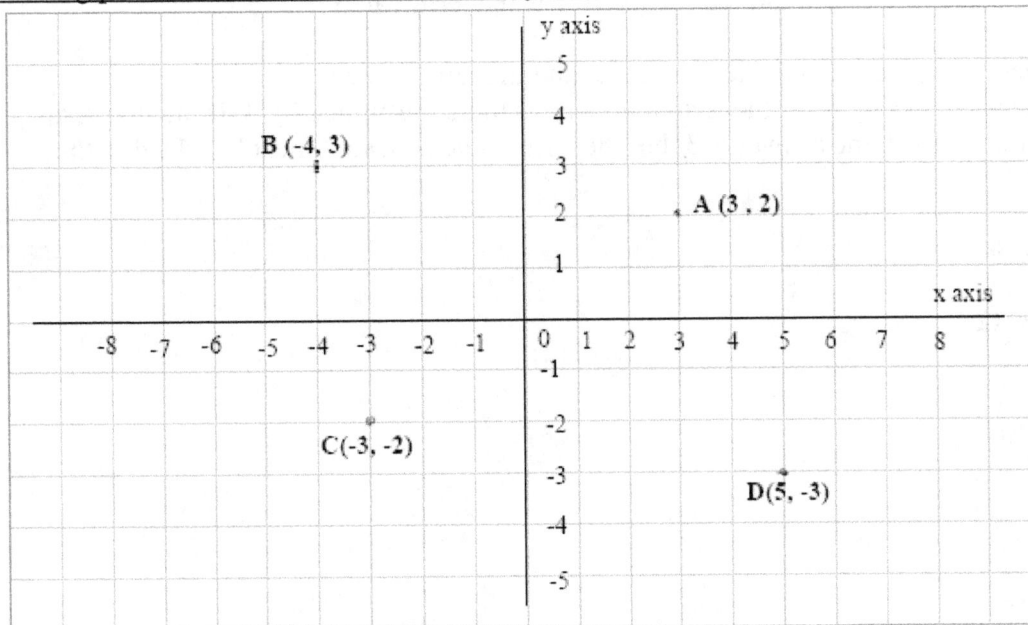

Page 50

The horizontal bar is the x axis and the vertical bar is the y axis.

Any number to the right of zero on the x axis is positive and to the left of zero is negative.

Similarly any number on the y axis above the x axis or above zero is positive, and any number below the x axis is negative.

Point A on the grid is **3** units to the right of **0** along the x axis and then **2** units up

3 and **2** are the ordered pair for point A.

The x value is always the first number in the ordered pair and the y value is the second.

Point B is **4** units to the left (**-4**) along the x axis and **3** units up (positive).

Move **3** units to the left of zero and then **2** units down to arrive at point C.

Practice 2

Plot the following points on a grid.
A **(6, 2)**   B (**-5, -4**) C (**-4, 6**)   D (**-5, 6**)   E (**-2, 5**)

Patterns

Determine the pattern for the sequence **6, 9, 12, 15**

6 is the first term, **9** is the second term, **12** is the third term etc.

This can be represented in a table.

Term number	1	2	3	4
Term value	6	9	12	15

The first row is the term number and the second row is the term value.

What is the relationship between the term number and the term value?

Since the term value goes up by **3**, the term number has to be multiplied by **3**. If the first term number is multiplied by **3**, the answer is **3**, but the term value is **6**, so **3** must be added to the answer to get 6.

$$3 \times 1 + 3 = 6$$
$$3 \times 2 + 3 = 9$$
$$3 \times 3 + 3 = 12$$

If we represent the term no by the variable n, then the relationship between the term number and the term value is

Term value = **3n + 3**

The equation for the term value is **3n + 3**

Example 1

Find the equation for the sequence **5, 11, 17, 23**. Find the value of the **50**th term.

Since the difference between the terms is **6**, the equation is **6n - 1** (multiplying the first term no by **6** will equal **6**, but the first term is **5,** so subtract **1** from **6**)

The equation becomes **6n – 1**   (**6 x 1 - 1 = 5**)

The **50**th term is **6 x 50 - 1 = 299**

Example 2

An equation to a pattern is **4n + 7**. The term value to the nth term is **171**. What term number does this correspond to?

$$171 = 4n + 7$$
$$171 - 7 = 4n$$
$$164 = 4n$$
$$\frac{164}{4} = n$$
$$n = 41$$

The term number of **41** corresponds to a term value of **171**.

Practice

1. Find the algebraic equation to the following sequences and find the value of the **30**th term.
   a) **1, 3, 5, 7**
   b) **10, 13, 16, 19**
   c) **15, 19, 23, 27**
   d) **25, 32, 39, 46**

2. A sequence starts with **5** and each term increases by **7**. What term number corresponds to a term value of **45**?

3. What term number corresponds to the last term in the following?
   a) **5, 11, 17, 23** --------- **71**
   b) **4, 12, 20** ----------**92**
   c) **8, 15, 22** -----------**76**

In the above examples a fixed amount was added to the previous term to get the current term. The sequences were linear as shown by the graph below.

The sequence **2 , 4, 6** means that the first term is **2**, the second term is **4** and the third term is **6**

Page 52

This can be plotted on the grid below with the term number along the *x* axis and the term value along the *y* axis.

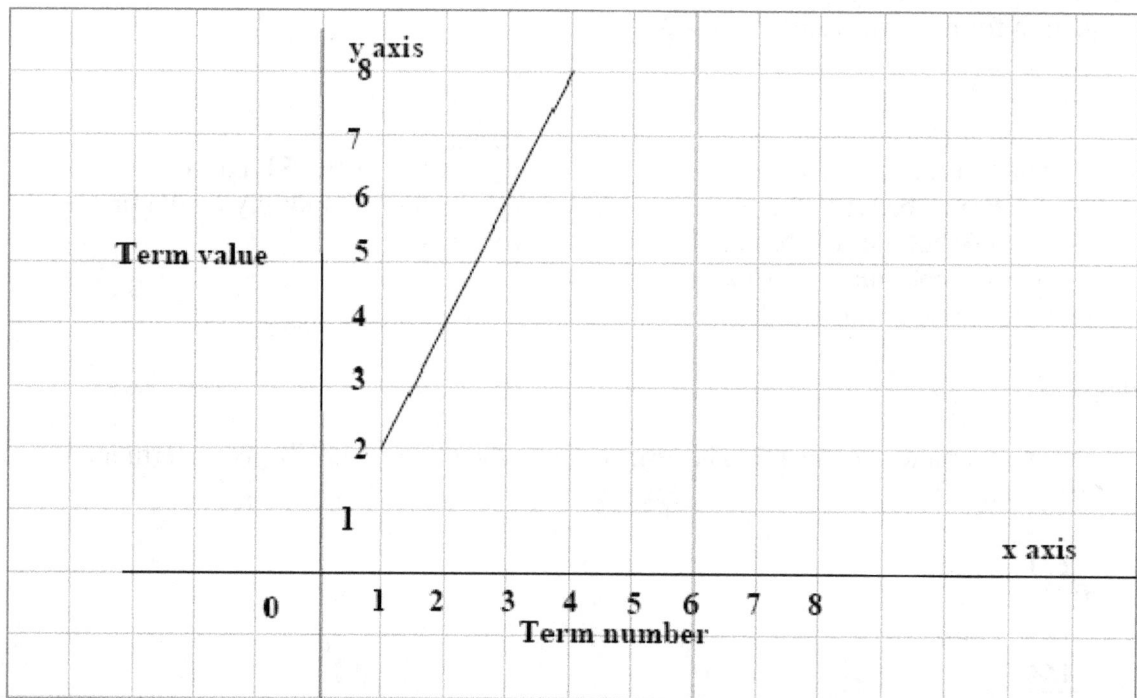

In the next set of sequences, the differences will not be the same but will be increasing.

Example 3

The rule to this pattern **1, 4, 9, 16**. is shown below.

$$1 + 3 = 4$$
$$4 + 5 = 9$$
$$9 + 7 = 16$$

The differences are **3, 5** and **7**   (4-1=3, 9-4=5, 16-9=7)

Term number	1	2	3	4
Term value	1	4	9	16

The general term for this sequence is n^2 {term value = (term no.$)^2$

$$1^2 = 1$$
$$2^2 = 4$$
$$3^2 = 9$$
$$4^2 = 16$$

Example 4

If the general term of a sequence is $n(n+1)$

Then the first term is $1(1+1)$ = **2**
The second term is $2(2+1)$ = **6**
The third term is $3(3+1)$ = **12**
The fourth term is $4(4+1)$ = **20**

The differences are **4, 6, 8** compared to **3, 5** and **7** in example **3**
Notice the differences are one more than example **3.**
4 is one more than **3**; **6** is one more than **5** etc
The general term for example **3** is $n(n)$ which is the same as n^2
The general term for example **4** is $n(n+1)$ and that is the reason why the differences are one more than those in example **3**.

Example 5

If the general term of a sequence is $n\,(n+2)$
Then the first term is **3**
The second term is **8**
The third term is **15**
The fourth term is **24**
The differences are **5, 7, 9** which is **2** more than the differences in example **3** and that is why the general term is $n(n+2)$

Example 6

The general term of a sequence is $(n+1)^2$
By substituting for n = **1, 2, 3** and **4**, the first four terms are **4, 9, 16, 25** compared to the first four terms of the sequence, n^2 which is **1, 4, 9, 16.**
The first term in the sequence $(n+1)^2$ starts with **4** so the sequence of $(n+1)^2$ is one ahead of the sequence of n^2
The general term of the sequence given by **9, 16, 25** would be $(n+2)^2$ because the sequence is two ahead of the sequence of n^2

By examining the terms of a sequence as in example **6** and the differences as in examples **3 ,4** and **5** , it is possible to know the general term of this type of sequence.

The pattern for **$n(n+2)$** is **3, 8, 15, 24**
The pattern for **$n(n+2)$** +**2** will be **2** more or **5, 10, 17, 26**. The differences are still the same **5, 7, 9** as for the pattern **$n(n+2)$**

Practice 4

Write an algebraic expression for the general term for the following.
a) **2, 5, 10 , 17**
b) **2, 6, 12, 20**
c) **4, 8, 14, 22**
d) **4, 9, 16, 25**
e) **6, 11, 18, 27**
f) **16, 25, 36**

Mean , mode and median

Math scores of different students	45	65	86	96	46	75	65	55	65	85	90	77	76	87	74	65

There are 16 students.
The mean of the scores is the total score divided by the number of scores.
Add all the scores and then divide the total by **16**
The total is equal to **1152**
Divide **1152** by **16**
= **72**
The mean or average score is **72**
The mode is the number that is repeated the most.
65 is the mode.
The median is the middle number after arranging the numbers in order.
The numbers arranged in order are **45, 46, 55, 65, 65, 65, 65, 74, 75, 76, 77, 85, 86, 87, 90, 96**
74 and **75** are the two middle numbers. Since the median can only be one number, add **74** and **75** and divide by **2**
74.5 is the median.

The same scores can be shown in steam and leaf .
The stem are the tens and the leaves are the units.

Stem	Leaves			
4	5	6		
5	5			
6	5	5	5	5
7	4	5	6	7
8	5	6	7	
9	0	6		

Example

a) What number must be added to the following data to increase the mean from **15** to **25**?
5, 7, 10, 15, 20, 23, 25
b) What number can be removed from the data and still keep the average at **15**?
c) What two numbers can be removed and the average would still be **15**?

Solution
a) The total of this set of **7** numbers is **105** which is the same as **15** x **7** as **15** is the mean. Since one number has to be added, the total number of numbers would have to be **8**. For **8** numbers with an average of **25** the total would have to be **25** x **8** = **200**. The number that has to added to the total of **105** is **200** -**105** = **95**. **95** has to be added after the number **25** to make the average of **25** for **8** numbers.

The same problem can also be solved with equations.

Let the number to be added be **x**.

$$\frac{105 + x}{8} = 25$$

(When the number is added to the total of **105** and divided by **8** then the average should be equal to **25**)

$$105 + x = (25)(8)$$
$$105 + x = 200$$
$$x = 200 - 105$$
$$x = 95$$

b) Since the average is **15**, then **15** can be removed and the average would still be **15**.

c) **20** and **10** can be removed and the average of the remaining **5** numbers would be **15**, since

$$\frac{20 + 10}{2} = 15$$

Practice

1. Find the mean, mode and median for the following data.
a) **23, 54, 40, 39, 54, 58, 67, 68**
b) **16, 35, 60, 57, 55, 87, 55, 65**
c) **34, 65, 67, 45, 36, 34, 56, 66**
d) **12, 24, 23, 34, 24, 37, 56, 46**

2. The mean for the following data is **25**.

   10, 15, 20, 25, 30, 35, 40
a) What number must be added to the set to make the mean **30**?
b) What number can be removed from the set to keep the mean at **25**?
c) What two numbers can be removed from the set to keep the mean at **25**?

3. Plot the stem and leaf for question **1**.   a, b , c and d combined.

CHAPTER 6

GEOMETRY 1

Angles

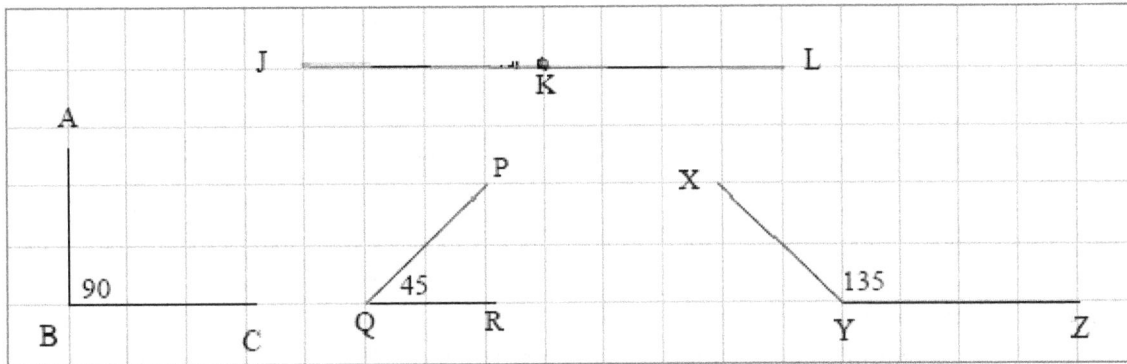

ABC is **90** degrees or a right angle. **PQR** is less than **90** degrees (acute angle). **XYZ** is more than **90** degrees but less than **180** degrees (obtuse angle). The straight line **JKL** is **180** degrees.

Lines

Lines that do not meet are parallel. **AB** and **CD** are parallel. Perpendicular lines are lines that intersect at right angles. **PQ** is perpendicular to **RS**.

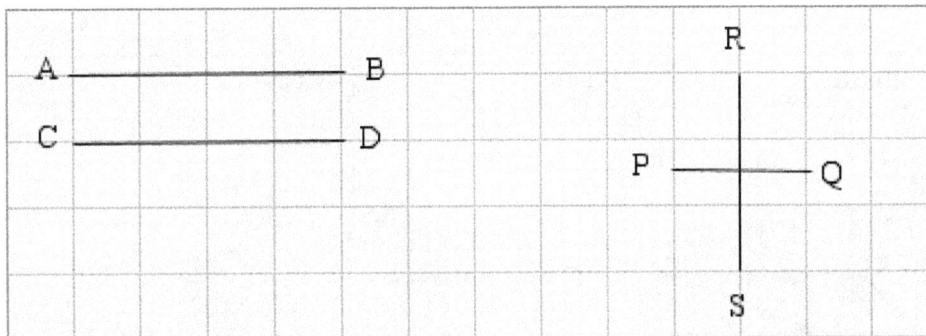

Triangles (refer to figure below)

Triangles are acute if all the angles are less than **90** degrees (ABC is an acute angled triangle).
An obtuse angled triangle is a triangle with one angle obtuse. (PQR is an obtuse angled triangle).
In a right angled triangle one angle is a right angle. (XYZ is a right angled triangle).
The sum of the angles in a triangle is **180°** . (symbol for degrees is °)

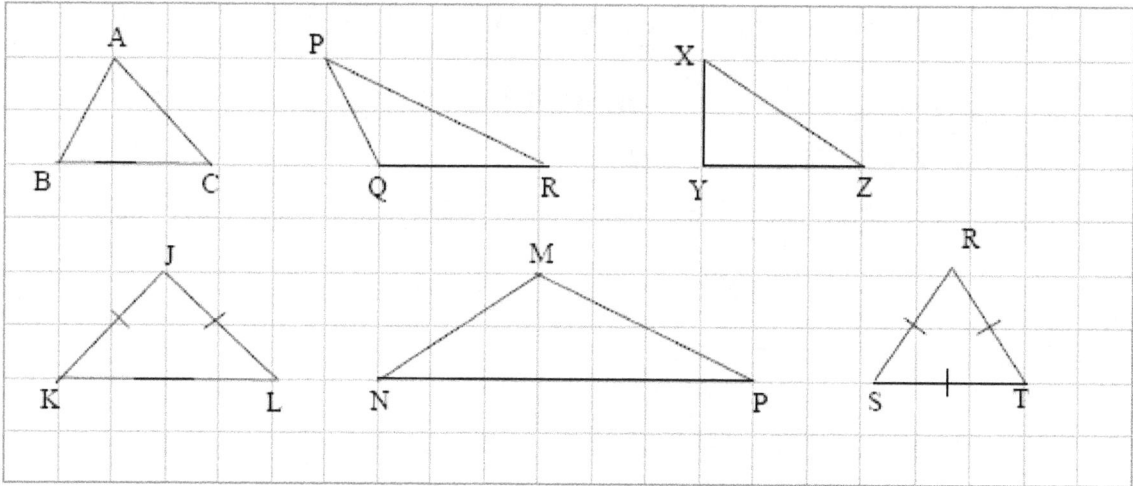

Triangles can also be classified by their sides.
- A triangle with two equal sides is called an isosceles triangle (**JKL** is an isosceles triangle **JK = JL**).
- A triangle with all sides of different lengths is scalene. (**MNP** is a scalene triangle)
- A triangle with all equal sides is an equilateral triangle. (**RST** is an equilateral triangle)

Properties of quadrilaterals

Quadrilaterals are four sided figures. *The sum of the angles in a quadrilateral is* **360°**

Fig 1

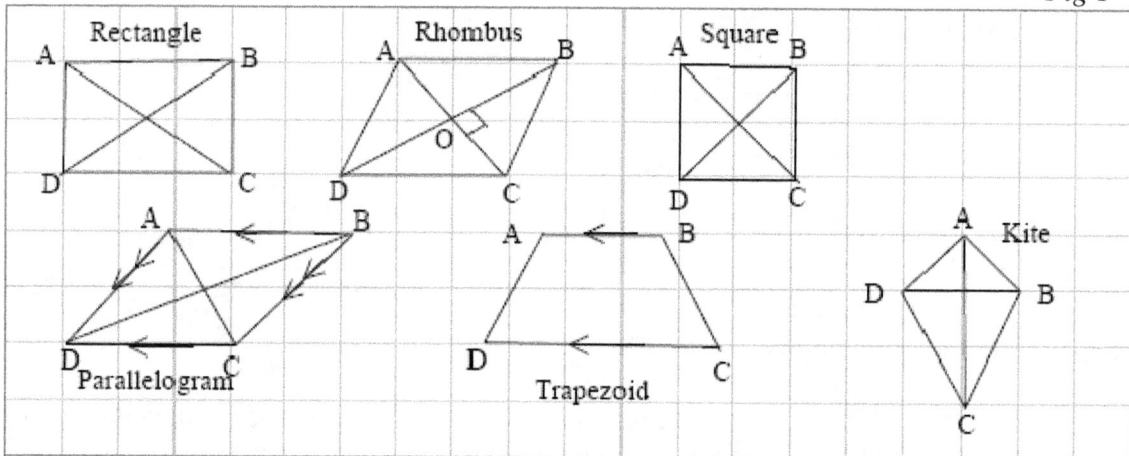

Rectangle (see Fig. 1)

All the angles in a rectangle are right angles. The opposite sides are equal. **AB = CD AD = BC**. The diagonals of a rectangle are equal and bisect each other. The diagonals of rectangle **ABCD** are **AC** and **BD**.

Page 59

Rhombus (see Fig. 1)

In a Rhombus, all four sides are equal **AB = BC = CD = DA**; the angles are not usually right angles. If the angles are all **90** degrees then the Rhombus is a square. The diagonals of a rhombus bisect each other at right angles, **AO = OC** and **BO = OD**. ∠BOC is a right angle (∠stands for angle) ∠AOB, ∠AOD, ∠DOC are all right angles.

Square (see Fig. 1)

A square has all sides equal and all angles are right angles. The diagonals of a square are equal and bisect each other at right angles.

Parallelogram (see Fig. 1)

In a parallelogram the opposite sides are equal and parallel. **AB** is equal and parallel to **DC**; **AD** is equal and parallel to **BC**. The diagonals of a parallelogram are not equal but bisect each other.

Trapezoid (see Fig. 1)

In a trapezoid one pair of opposite sides are parallel. **AB is** parallel to **DC**.

Kite (see Fig. 1)

In a kite, two pairs of adjacent sides are equal. **AD** and **AB** are adjacent sides because they are touching each other. **CD** and **CB** are another pair of adjacent sides. **AD = AB** and **CD = CB**. The diagonals of a kite are perpendicular to each other.

Practice 1

Name the quadrilateral or quadrilaterals with the following properties of the diagonals below.
a) The diagonals bisect each other at right angles.
b) The diagonals are equal.
c) The diagonals are perpendicular to each other.
d) The diagonals are not equal.

Area

A, **B** and **C** are the vertices of triangle **ABC**.

The area of triangle **ABC** is $\dfrac{\textbf{base x height}}{\textbf{2}}$

The height of a triangle is measured from a perpendicular drawn from a vertex to the opposite side or opposite side extended.

In the Fig. 2, the height **AD** (h) of triangle **ABC** is measured from vertex **A** to **BC** (**BC** is the base) or a perpendicular from vertex **B** to side **AC** (**AC** is the base) or a perpendicular drawn from **C** to **AB** in which case **AB** would be the base.

In triangle **PQR**, the height is measured from **P** to **RQ** extended. The area of this triangle is equal to $\dfrac{\text{h x RQ}}{2}$

The area of triangle **LMN** is $\dfrac{\text{h x LM}}{2}$ (**LM** is the base)

There are three ways to measure the height of any triangle because there are three vertices. Each method will result in the same area for a given triangle.

Fig 2

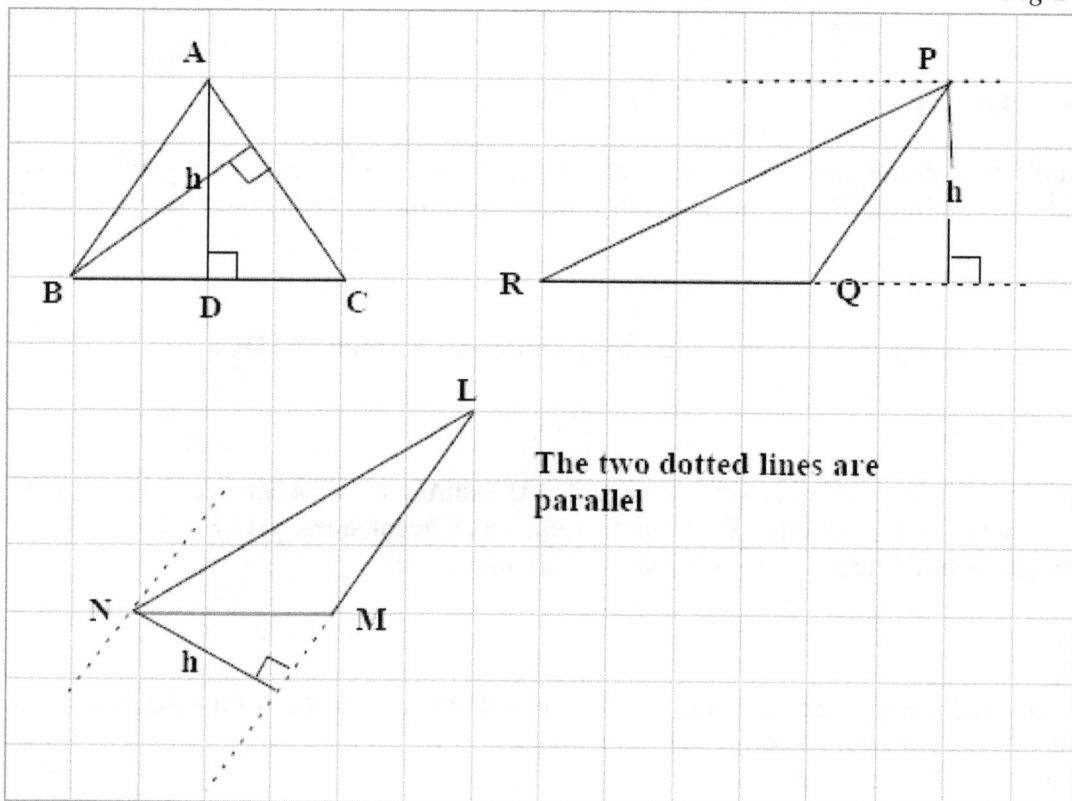

The two dotted lines are parallel

The area of a parallelogram is equal to the base x height (Fig. 3). The base is **DC** and the height is *h* which is the distance between **AB** and **DC**. (*h* is perpendicular to **AB** and **DC**). The height can also be measured by dropping a perpendicular from vertex **B** to **DC** extended.

The area of a trapezoid is half the sum of the parallel sides *x* the distance between them.

The area of trapezoid is $\dfrac{(\textbf{AB} + \textbf{DC})\textbf{h}}{2}$ (Fig. 3)

The distance between two parallel lines is always the shortest distance between them, which is the line that is perpendicular to both parallel lines.

The distance between **PQ** and **RS** is *h* and not *b*. (Fig. 3).

Fig 3

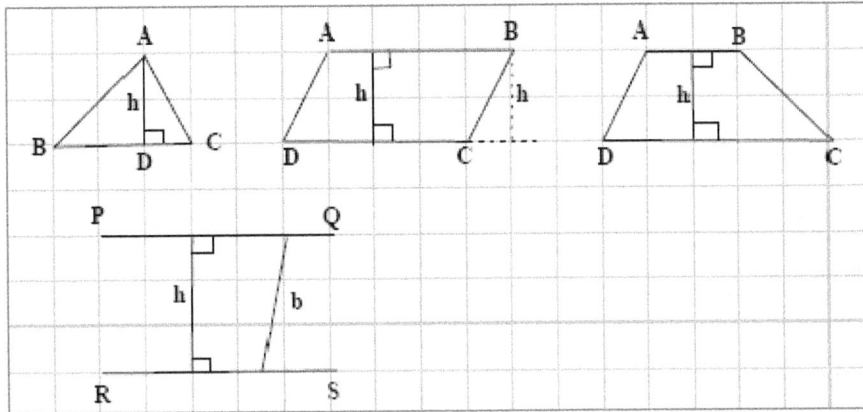

The perimeter of a figure is the sum of all its sides.
The perimeter of triangle **ABC = AB + BC + CA**.

Example

Calculate the area and perimeter of triangle **ABC** in Fig. 4.

The area of triangle ABC = $\dfrac{AD(BC)}{2}$ = $\dfrac{3(2)}{2}$ = **3 cm^2**

The perimeter is **2.5 + 3 + 3.6 = 9.1** cm

Calculate the area and perimeter of parallelogram **ABCD** (Fig. 4)
The area of parallelogram **ABCD = DC** (*h*) = 4(2.5) = **10 cm^2**
The perimeter of parallelogram **ABCD** is **4x2 +3x2 = 14** cm

Calculate the area and perimeter of the trapezoid **ABCD** (fig **4**)
The area of the trapezoid is $\dfrac{(AB+DC)h}{2}$ = $\dfrac{(2 + 5)(2)}{2}$ = $\dfrac{(7)(2)}{2}$ = **7 cm^2**
The perimeter of trapezoid **ABCD** is **2.5 +2 + 3 + 5 = 12.5** cm

Practice 2

Find the areas to the following figures:

a) A triangle with a base of **4** cm and a height of **5** cm.
b) A parallelogram with a base of **6** cm and a height of **7** cm.
c) A trapezoid with one parallel side equal to **5** cm and the other parallel side equal to **4** cm.
    The height of the trapezoid is **6** cm.

Examples of length and perimeter increases
If the width of a rectangle **PQRS** in Fig. 4, stays the same and the perimeter is increased by **10** cm, what is the increase in length of the rectangle?

Since the perimeter is increased by **10** cm, each length is increased by **5** cm as there are two lengths. The length of the rectangle is increased by **5** cm.

Fig 4

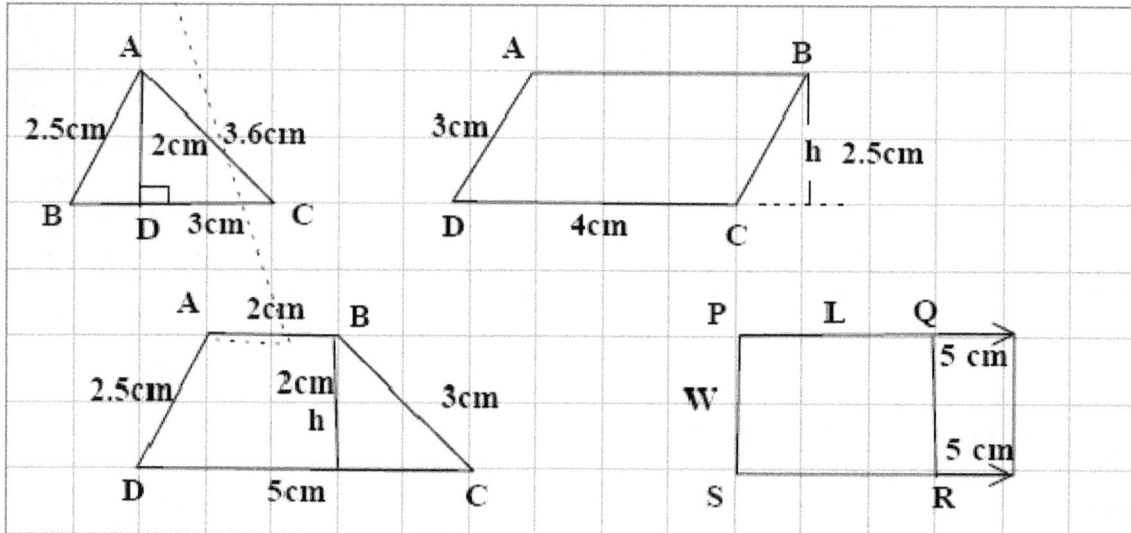

If the length of a rectangle is increased by **20** cm and the width remains the same, by how much has the perimeter increased?

Two lengths are increased by **20** cm, so the perimeter has increased by **40** cm.

Practice

3. If the perimeter of an equilateral triangle is increased by **30** cm, by how much has each side increased?

4. If the side of equilateral triangle is increased by **5** cm, by how much has the perimeter increased?

5. If the perimeter of a rectangle is increased by **30** cm and the length stays the same, by how much has the width increased?

6. If the length of a rectangle is increased by **20** cm and the width stays the same, by how much has the perimeter increased?

Polygons are classified according to the number of sides

A **3** sided figure is a triangle.
A **4** sided figure is a quadrilateral.
A **5** sided figure is a Pentagon.
A **6** sided figure is a Hexagon.
A **7** sided figure is a Heptagon.

An **8** sided figure is an Octagon.
A **9** sided figure is a nonagon.
A **10** sided figure is a decagon.

If the sides are all of equal length then the figure is called regular, e.g. a **5** sided figure with all the sides equal is called a regular pentagon.

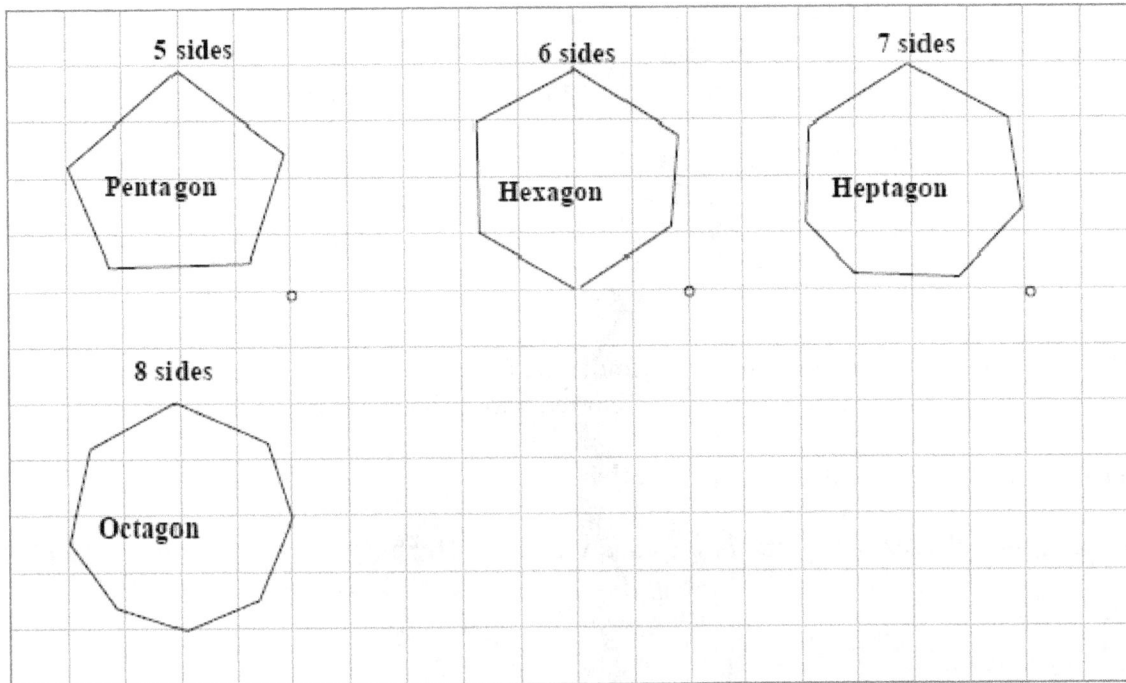

5 sides — Pentagon
6 sides — Hexagon
7 sides — Heptagon
8 sides — Octagon

Circles

The circumference is the outline of the circle. (See figure on page 65) The radius (r) is the distance from the centre of the circle to the circumference of a circle.

The length of the circumference is **2Πr.** (Π = **3.14** approximately). **C = 2Πr**
The diameter is twice the radius. The diameter is a straight line passing through the centre of the circle. The circumference in terms of the diameter is **C = Πd**.

The area of a circle is **Πr²**

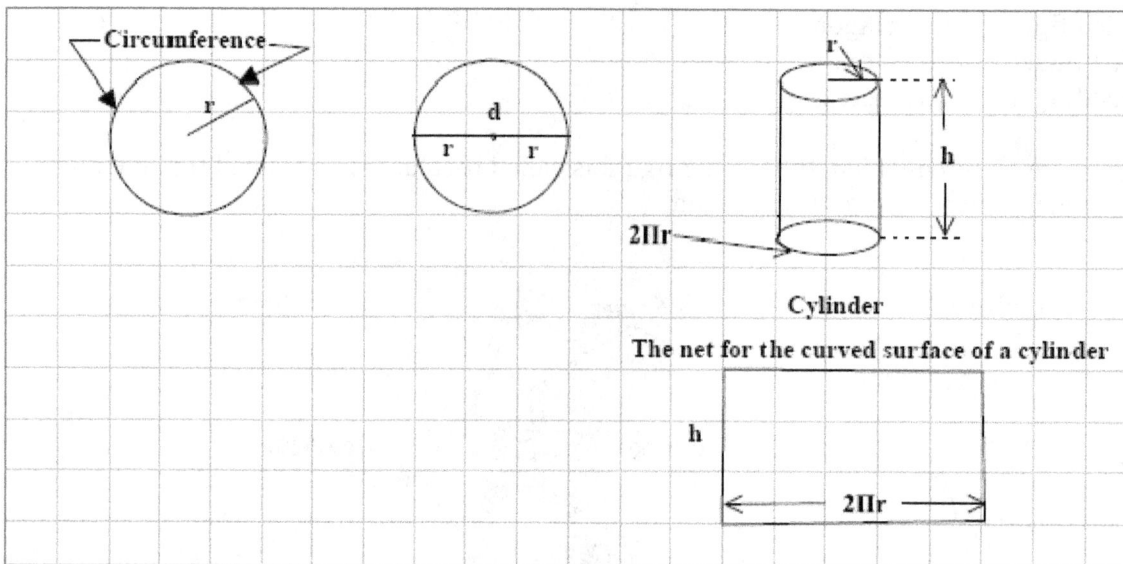

The surface area of the top and bottom of a cylinder is $2\Pi r^2$ (the surface area of the top circle and the bottom circle). The surface area of the curved part is $2\Pi rh$.

The total surface area of a cylinder is $2\Pi r^2 + 2\Pi rh$

The volume of the cylinder = the area of the base x height = $\Pi r^2 h$ (the area of the base is Πr^2).

Example

A cylinder has a radius of **3** cm and a height of **4** cm. Find the surface area of the cylinder and the volume.

The surface area of a cylinder is $2\Pi r^2 + 2\Pi rh$
$$= 2\Pi(3^2) + 2\Pi(3)(4)$$
$$= 6.28(9) + 6.28(12)$$
$$= 56.52 + 75.36$$
$$= 131.88 \text{ cm}^2$$

The volume of the cylinder is $\Pi r^2 h = \Pi(3^2)(4)$
$$= \Pi(9)(4)$$
$$= 113.1 \text{ cm}^3$$
The volume of the cylinder is **113.1 cm³**

Example

Find the area of the unshaded portion in Fig. 2.

The area of the larger circle is $\Pi(6^2)$.
The area of the smaller circle is $\Pi(2^2)$.

The area of the unshaded portion is the difference of the two areas.
Area of the unshaded portion is $\Pi(36) - \Pi(4)$
= 113.10 – 12.57
= 100.53 cm^2

Fig 2

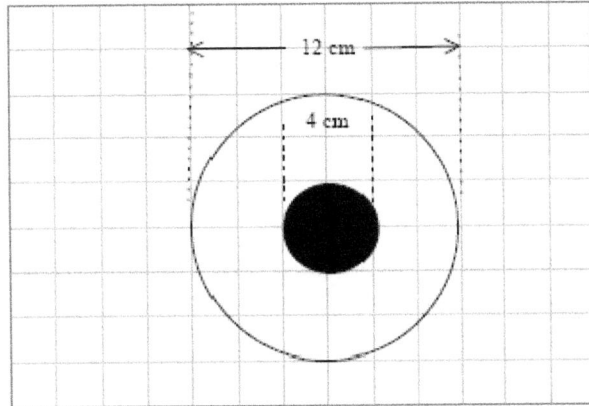

Pythagorean theorem

The Pythagorean theorem states that in a right angled triangle, the square on the hypotenuse is equal to the sum of squares of the other two sides.

Fig 1

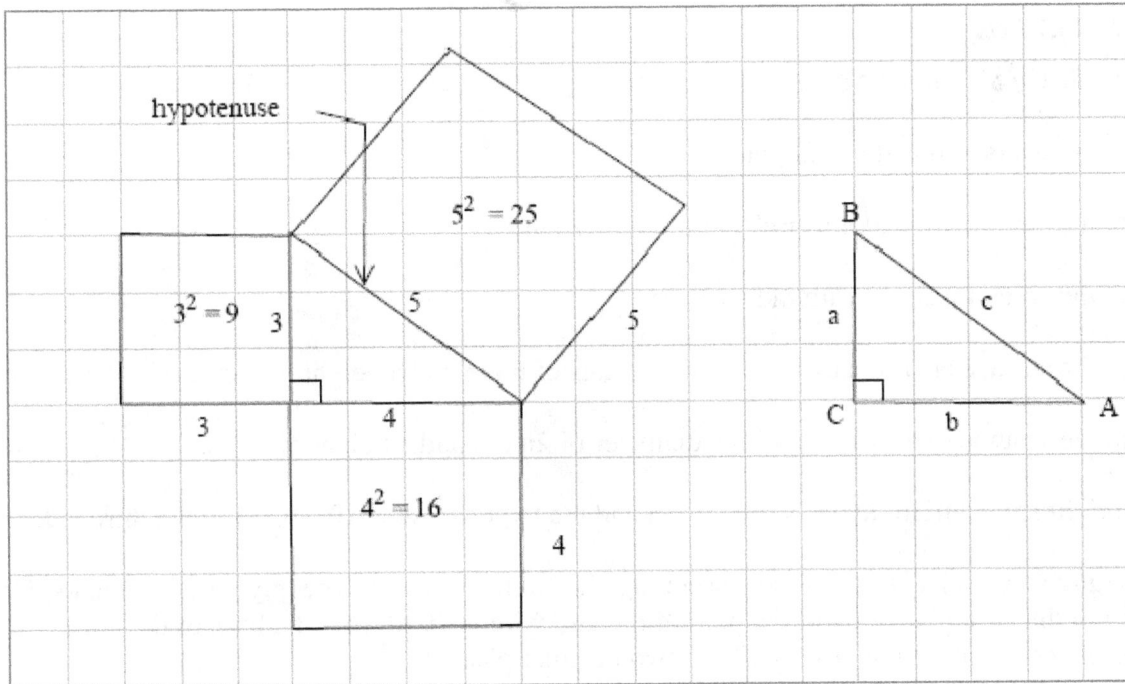

The hypotenuse is the longest side or the side opposite the right angle.
The square on the hypotenuse = sum of the squares of the other two sides.

$$5^2 = 4^2 + 3^2$$
$$25 = 16 + 9$$

In triangle **ABC** Fig.1, the sides opposite the vertices are in small letters and they represent the sides. The side opposite vertex **A** is marked as *a,* which is the length of **BC.**

Using the Pythagorean theorem $c^2 = a^2 + b^2$, if any two sides are given, the third can be found by rearranging the equation. In an equation, the left hand side of the equal sign is equal to the right hand side of the equal sign.

$$c^2 = a^2 + b^2 \text{ or}$$
$$a^2 + b^2 = c^2$$
$$a^2 = c^2 - b^2 \quad (b^2 \textit{ was on the left hand side but is now on right hand side of the equal}$$
$$\textit{sign, so the sign changes to minus}).$$

Similarly, $b^2 = c^2 - a^2$.

In Fig. 1, if $a = 4$ cm and $c = 7$ cm find b.
$$b^2 = c^2 - a^2$$
$$b^2 = 7^2 - a^2$$
$$b^2 = 49 - 16$$
$$b^2 = 33 \qquad \textit{(taking square roots of both sides refer to the end of chapter 4 on measurements)}$$
$$\sqrt{b^2} = \sqrt{33}$$
$$b = 5.74 \text{ cm}$$

To prove that $\sqrt{b^2} = b$, take a numerical example $\sqrt{5^2} = \sqrt{25} = 5$; $\sqrt{5^2} = 5$

Practice (answers to one decimal place)

7. Find the circumference of a circle with radius **7** cm.

8. Find the area of a circle with radius **5** cm.

9. Find the surface area of a cylinder with a radius of **6** cm and a height of **4** cm.

10. Find the volume of a cylinder with a diameter of **20** cm and a height of **6** cm.

11. In a right angled triangle one side is **6** cm and the hypotenuse is **10** cm. Find the third side.

12. One guy wire on an electrical pole **60** m high has to be replaced. The guy wire is attached **50** m from the ground to support the pole. If the guy wire is **30** m from the base of the pole, what is the length of the guy wire? (to two decimal places)

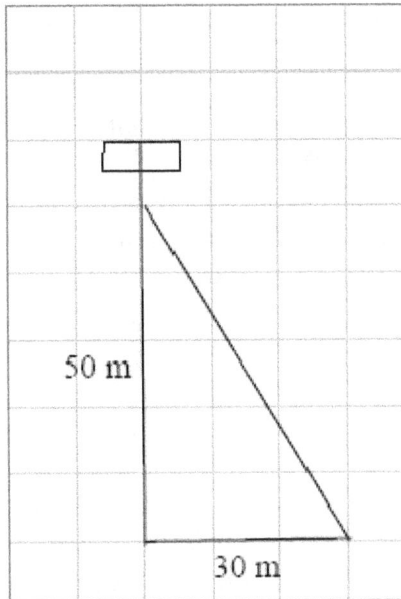

13. The perimeter of a regular hexagon is **60** cm. What is the length of the side?

The volume and surface area of a rectangular prism

Fig 1

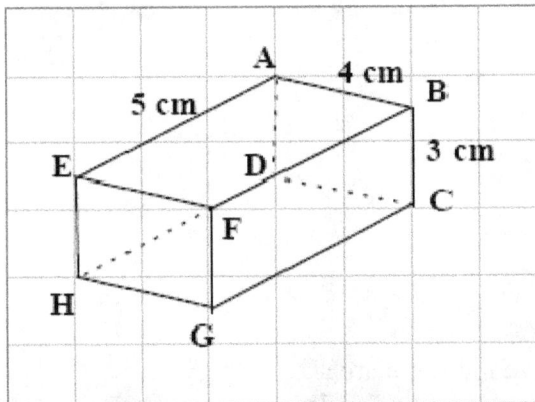

To calculate the surface area of this rectangular prism, notice there are three sets of equal parallel faces

These are **AEHD** and **BFGC** (Two side faces)
EFGH and **ABCD** (Front and back face)
ABFE and **DCGH** (top and bottom face)
If **AE = 5**cm; **AB = 4** cm and **BC = 3** cm, find the surface area of the prism.
The area of the side face **AEHD = AE x AD = 5 x 3 (AD = BC = 3** cm **) = 15 cm^2**

The area of the two side faces = **15 x 2 = 30** (since the two side faces are equal in area)
The area of the front face **EFGH = EF x FG** (**EF = AB = 4** cm and **FG = BC = 3** cm)

The area of the front face = **4** x **3** = **12** cm^2 which is also equal to the area of the back face.
<u>The area of the front and back face = **12** x **2** = **24** cm^2</u>
The area of the top face = **ABFE** = **AB** x **BF** (**AB** = **4** cm; **BF** = **AE** = *5* cm)
The area of the top face = **4** x **5** = **20** cm^2 which is equal to the area of the bottom face.
<u>The area of the top and bottom face = **20** x **2** = **40** cm^2</u>
The total surface area = the area of the two side faces = **30** cm^2
Plus the surface area of the front and back face = **24** cm^2
Plus the area of the top and bottom face = **40** cm^2
= **30 + 24 + 40 = 94** cm^2

The volume of Fig 1 = length x width x height
= **5** x **4** x **3** = **60** cm^3

Fig 2

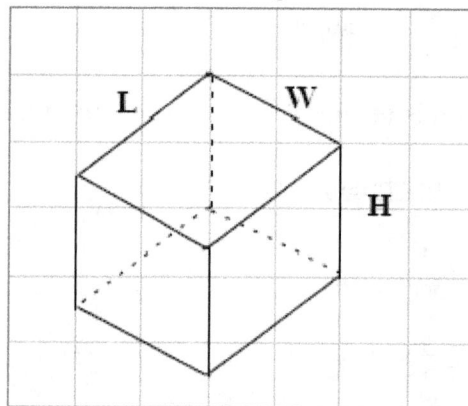

The surface area of Fig *2*
      = **L** x **W**   (top and bottom face)
      + **W** x **H**   (front and back face
      + **H** x **L**   (side faces)

Since there are two of each face that have the same area.
The total surface area = **2LW + 2WH + 2HL** (2 x L x W means as **2LW**) or 2(**LW + WH + HL**)

To prove that the two are the same, take a numerical example.
2 x 3 + 2 x 4 + 2 x 5 = 6 + 8 + 10 = 24
2(3 + 4 + 5) = 2(12) = 24 (Whatever is in the brackets are added first and then the result is multiplied by the number outside the bracket.) **2(4)** means **2** multiplied by **4**.
To remember the formula for the surface area of a rectangular prism, write **L— W— H** and then take two at a time. First take **LW** then **WH** and finally **HL.**

Example

A rectangular prism is **6** cm high, **8** cm long and **4** cm wide.
Find the surface area of the prism.
The surface area = **2(LW + WH + HL)**
= **2(8 x 4 + 4 x 6 + 6 x 8)**
= **2(32 + 24 + 48)**
=**2(104)**
= **208** cm^2
The volume of a rectangular prism = **L x W x H**
The volume of the prism = **8 x 4 x 6 = 192** cm^3
Another way of looking at the volume of this prism is the area of the base x height.
The area of the base is **L x W** and the height is **H**.

Practice

14. Calculate the surface area of the following rectangular prisms.
  a) Height = **7** cm ; width = **5** cm ; length = **6** cm
  b) Width = **8** cm ; height = **6** cm and length = **9** cm
  c) Length = **11** cm ; width = **5** cm and height = **7** cm

The volume and surface area of a triangular prism

In a triangular prism two faces are parallel and congruent. The prism in Fig. 3 has two parallel and congruent faces. Triangle **ACB** and triangle **DFE** are parallel to each other and congruent. The base of the prism is always one of the congruent faces. In this case, Δ**DFE** is the base.

Example

Find the volume and surface area of the triangular prism in Fig. 1

Fig. 1

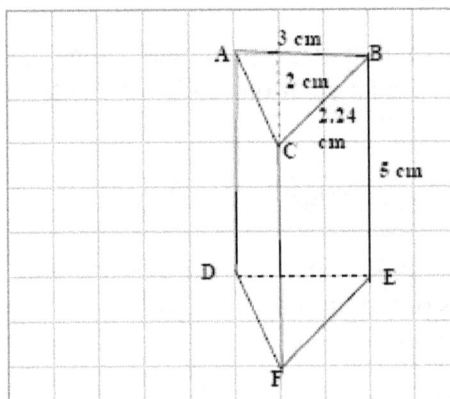

The volume of the triangular prism is the area of the base x height (Fig.1). The area of the base is the area of the triangle with base = **3** cm and height = **2** cm.

The area of the triangle = $\dfrac{\textbf{base x height}}{\textbf{2}}$ = $\dfrac{\textbf{3 x2}}{\textbf{2}}$ = **3** cm^2

The height of the triangular prism is **5** cm. The volume of the prism **3** x **5** = **15** cm^3

The surface area of the two rectangular faces **ADFC** and **BEFC** = **5** x **2.24**

$\qquad\qquad\qquad\qquad\qquad\qquad\qquad\qquad\qquad\qquad$ = **11.2** cm^2 x **2**

$\qquad\qquad\qquad\qquad\qquad\qquad\qquad\qquad\qquad\qquad$ = **22.4** cm^2 (Fig. 1)

The area of rectangle **ABED** is **5** x **3** = **15** cm^2

The area of the two triangular faces is **ACB** and **DEF** = **2** $\left(\dfrac{\textbf{3 x2}}{\textbf{2}}\right)$ = **6** cm^2

The total **SA** of the triangular prism is **22.4** + **15** + **6** = **43.4** cm^2

Note

If the prism was resting on a rectangular face, **ACDE** as in Fig. 2, then the calculation based on area of the base x height would not work. For the calculation to work the base and the top must be the same (congruent). In this case the base is a rectangle and the top is an edge.

Fig. 2

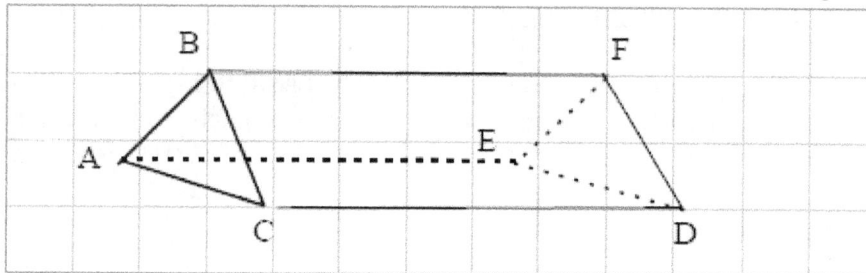

ACDE is the base and **BF** is the edge. The volume for this triangular prism is equal to the area of triangle **ABC** multiplied by the length **CD**.

With a rectangular prism, it does not matter which face is the base because the top face will be congruent to the bottom face, so the calculation works no matter how the rectangle rests (all opposite faces are congruent).

Expressing formulas in terms of a variable

In any formula, if one variable is unknown, the unknown variable can be calculated by expressing the formula in terms of the unknown variable.

Example

To express h in terms of r and h, in the formula for the volume of a cylinder, $\textbf{V} = \textbf{\Pi r}^2\textbf{h}$
$\textbf{\Pi r}^2\textbf{h} = \textbf{V}$. ($\Pi r^2$ is the multiplier of h on the **LHS** of the equation, becomes the divisor of V or

the denominator on the **RHS** of the equation).

$$h = \frac{V}{\Pi r^2}$$

or to express **r** in terms of **V** and **h.**

$$\Pi r^2 h = V$$

$$r^2 = \frac{V}{\pi h}$$

(Taking the square root of both sides)

$$\sqrt{r^2} = \sqrt{\frac{V}{\pi h}}$$

$$r = \sqrt{\frac{V}{\pi h}}$$

Example

The formula for the volume of a rectangular prism is equal to **L x W x H.**

$$LWH = V$$

To express **L** in term of **V**, **W** and **H**: $\quad L = \frac{V}{WH}$

To express **W** in terms of **V**, **L** and **H**: $\quad W = \frac{V}{LH}$

To express **H** in terms of **V**, **L** and **W**: $\quad H = \frac{V}{LW}$

Example

If the volume of a rectangular prism is **500 cm^3**, the length is **15** cm and the height is **10** cm, calculate the width.

$$W = \frac{V}{LH}$$

$$\text{Width} = \frac{500}{15 \text{ x } 10} = \frac{500}{150} = 3.33\text{cm}$$

Example

The formula for the area of a triangle is $\frac{b \text{ x } h}{2}$

$$A = \frac{b \text{ x } h}{2}$$

$$2A = b \text{ x } h \quad \text{or } b \text{ x } h = 2A$$

$$b = \frac{2A}{h} \; ; \; h = \frac{2A}{b}$$

If the area and base of a triangle were given, then the height can be calculated. If the area and height were given, then the base can be calculated.

Review

1. Calculate the area of the following figures.
   a) A triangle has a base of **5** cm and a height of **4** cm.
   b) A parallelogram has a base of **6** cm and a height of **50** mm (calculate area in cm)
   c) A trapezoid has one parallel side equal to **5** cm and the other parallel side equal to **4** cm. The height of the trapezoid is **3** cm.
   d) Calculate the circumference and area of a circle with radius **5** cm.
   e) Calculate the surface area of a cylinder with a radius of **5** cm and a height of **4** cm.

2. Calculate the unshaded areas of the figures in fig **3** marked **A**, **B**, **C** and **D**.

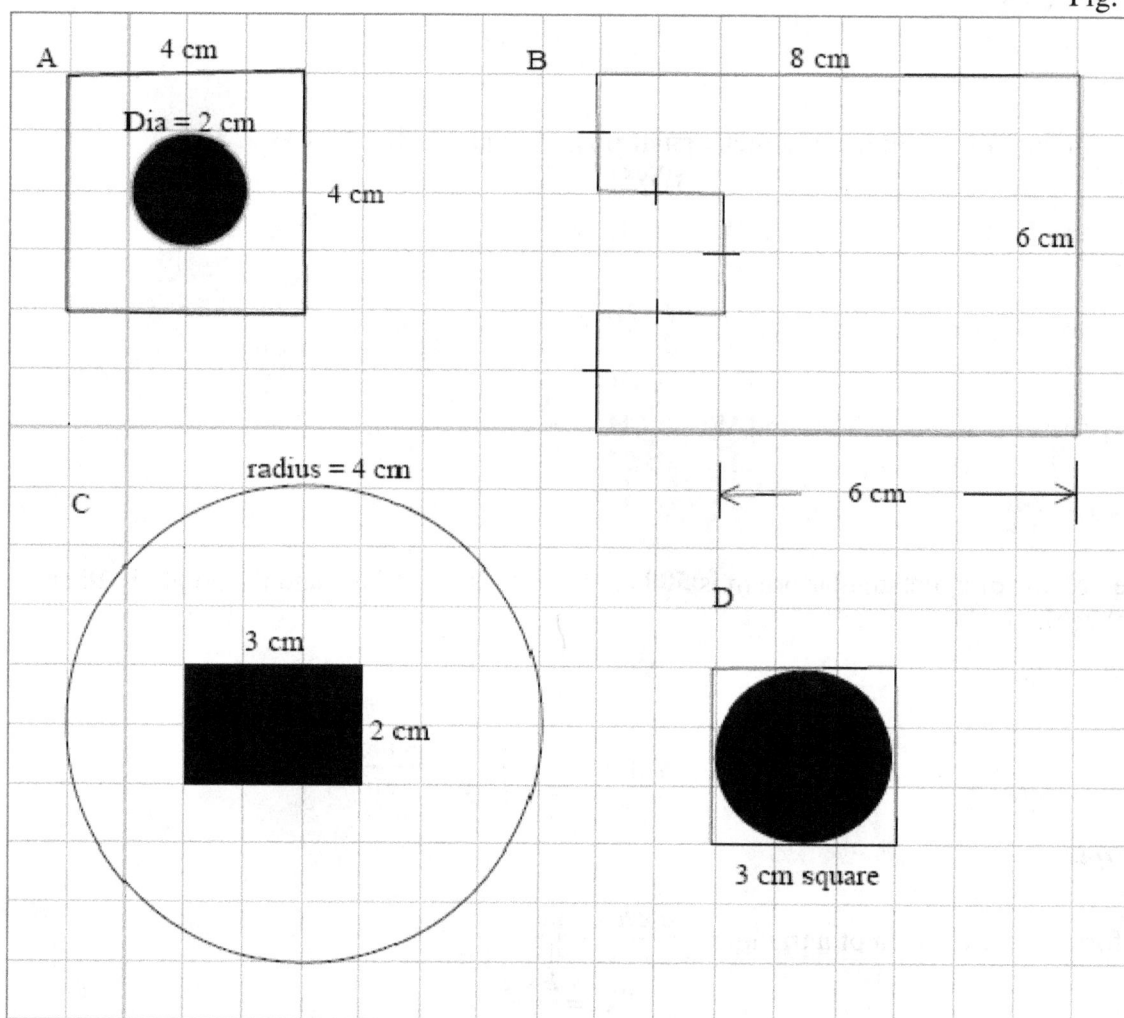

Fig. 3

3. a) Find the area of triangle **ABC** in fig **4** and the areas of the three triangles below triangle **ABC**.
   b) What fraction is triangle **ABC** of parallelogram **PQBC**?

Fig 4

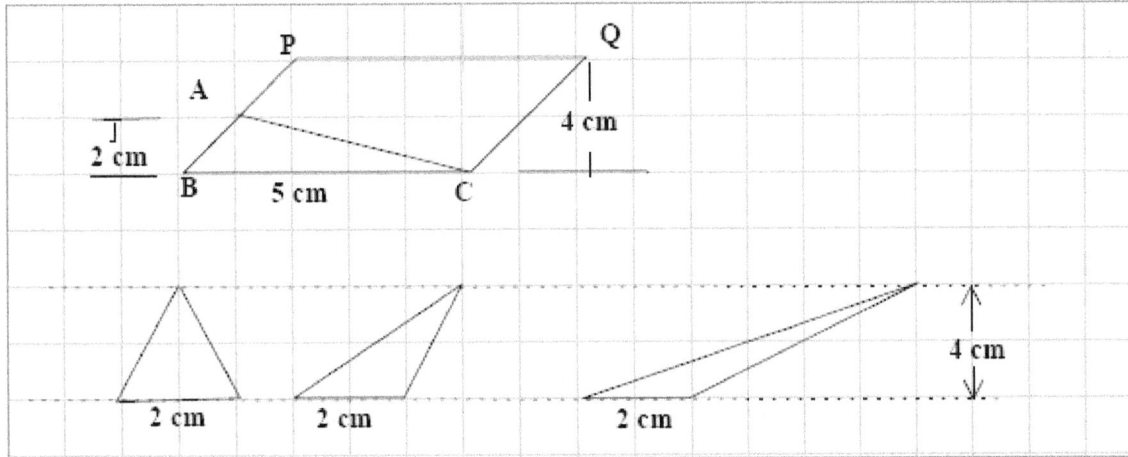

4. The area of a triangle is **30** cm^2. The base is **6** cm, find the height?

5. The area of a parallelogram is **60** cm^2. If the base is **10** cm, find the height of the parallelogram.

6. The area of a trapezoid is **40** cm^2. The height is **5** cm. Find the sum of the parallel sides.

7. If the sum of the parallel sides in a trapezoid is **20** cm and the area is **60** cm, find the height.

8.
   a) Calculate the surface area and the volume of a cylinder with a radius of **4** cm and a height of **5** cm.
   b) Find the surface area and volume of a rectangular prism with a length of **5** cm a width of **4** cm and a height of **3** cm.
   c) Calculate the surface area and volume of the figures in Fig 5. marked **A** and **B**.

Fig 5

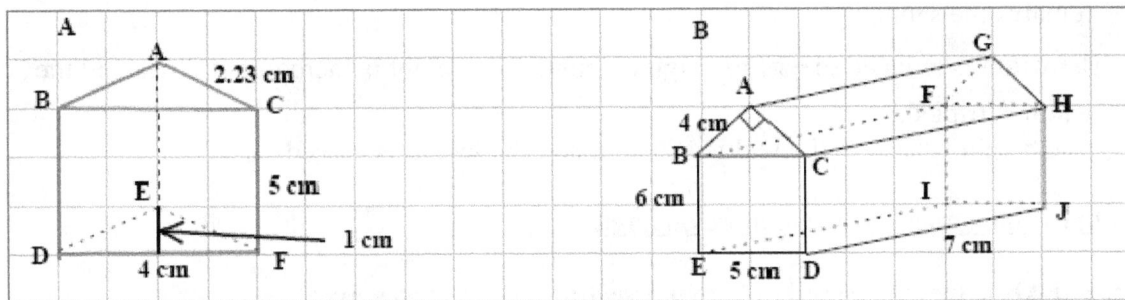

9. A pizza box is **30** cm by **30** cm. Find the area of the largest pizza that can fit in the box (pizza is circular).

10. If the hypotenuse of a right angles triangle is **5.83** cm and one side is **5** cm, what is the length of the other side?

11. One side of a right angles triangle is **4**, the hypotenuse is **5**. Find the third side.

<u>Relationship between length, width, height and volume, increases or decreases of rectangular prisms</u>

Example

If the length of a rectangular prism is **5** cm, the width is **4** cm and the height is **2** cm, the volume = **5** x **4** x **2**.

> The volume = **40** cm^3.
> If the length is doubled to **10** instead of **5** cm, the volume = **10** x **4** x **2** = **80** cm^3.
> The previous volume was **40** cm^3, but the new volume is **80** cm^3.
> The volume has doubled.

If the length is doubled, but the height and width remain the same, then the volume is doubled. If the width is doubled, but the length and height stay the same, then the volume is doubled. If the height is doubled and the other two dimensions remain the same, then the volume is doubled.

If the length is doubled and the width is doubled then the new volume will be **2** x **2** = **4** times the original volume.

If the length is tripled, the width is doubled and the height is tripled then the new volume will be **3** x **2** x **3** = **18** times the original volume.

12. Find the new volume in each of the following rectangular prisms, if the original volume is **50** cm^3.
    a) The length is doubled, the height is halved and the width is halved.
    b) The length is tripled, the width is halved and the height is halved.
    c) The height is reduced to $\frac{1}{3}$ of its original height, the width is tripled and the length remains the same.
    d) The width is reduced to $\frac{1}{4}$ of its original length, the height is increased **4** times and the width is tripled.
    e) The height is halved, the width is halved and the length is halved.

<u>Parallel lines and a transversal and more on angles</u>

PQ||RS and **AB** is the transversal (a line intersecting two or mores lines).

Fig 6

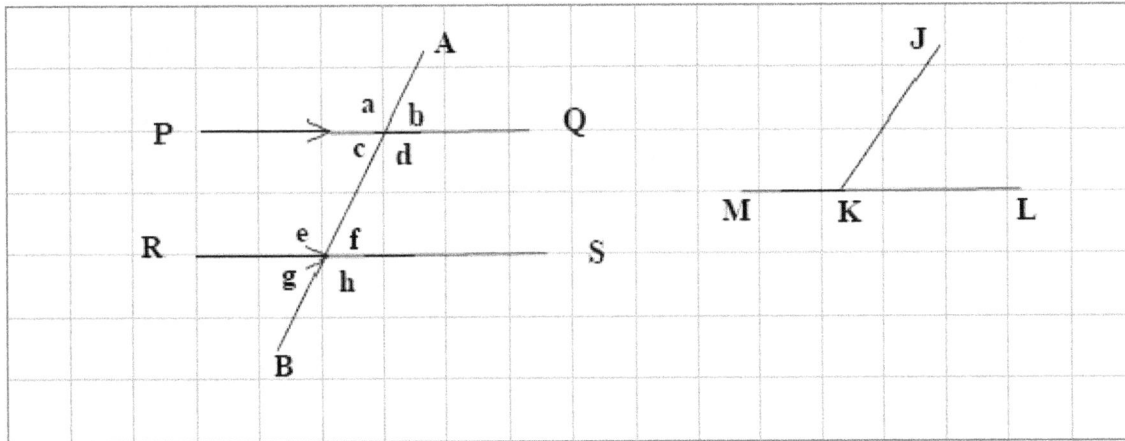

Refer to Fig. 6.

$\angle b = \angle f$, $\angle d = \angle h$, $\angle a = \angle e$, $\angle c = \angle g$ (corresponding angles)
$\angle c = \angle f$, $\angle d = \angle e$ (alternate angles)
$\angle a = \angle d$, $\angle b = \angle c$; $\angle e = \angle h$, $\angle f = \angle g$ (vertically opposite angles)
$\angle d + \angle f = 180°$; $\angle c + \angle e = 180°$ (interior angles on the same side of the transversal)

In Fig. 6 to the right $\angle JKL + \angle JKM = 180°$ (**180°** is the symbol for **180** degrees)

Fig 7

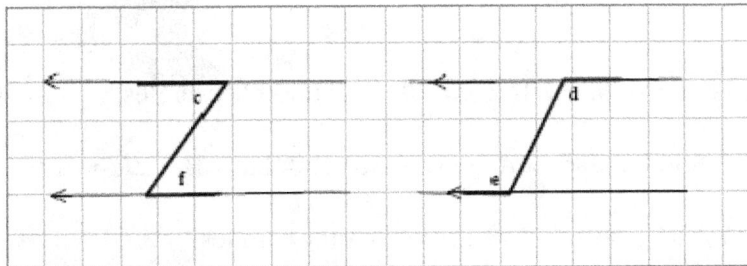

In Fig.7 the alternate angles *c* and *f*; *d* and *e* are shown in the form of the letter **Z**.

Fig 8

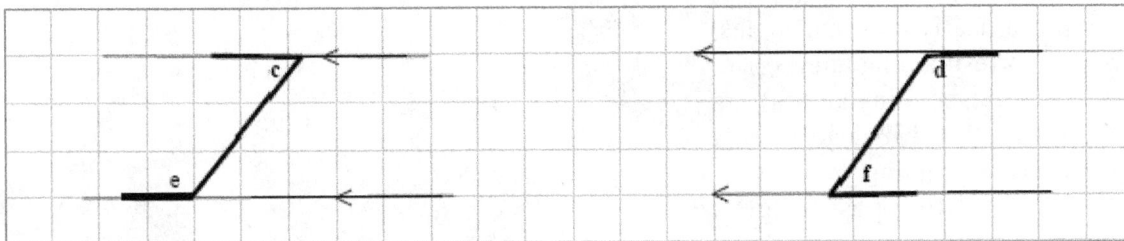

In Fig. 8, the interior angles on the same side of the transversal *c* and *e*; *d* and *f* are shown in the form of the letter **C**.

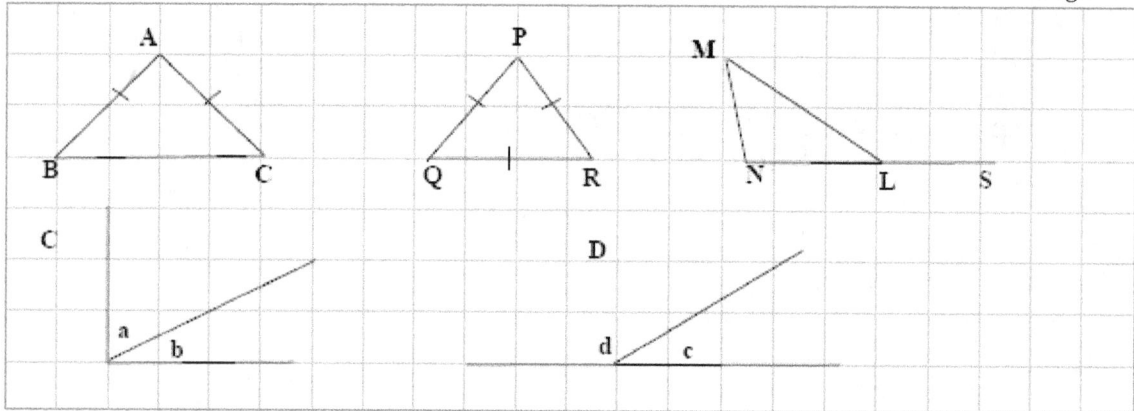

Fig 9

In Fig. 9, △**ABC** is isosceles **AB = AC**; ∠**B** = ∠**C**. (The base angles in an isosceles △ are equal) **PQR** is equilateral ∠**P** = ∠**Q** = ∠**R** = **60°**.

In △**MNL**, the exterior angle **MLS** is equal to the sum of the interior opposite angles **M** and **N**. If **MLS = 130°** then **M + N = 130°**.

In Fig. 9 **C**, *a* and *b* are complementary angles because their sum is **90°**.

In Fig. 9 **D**, *c* and *d* are supplementary angles as their sum is **180°**.

Example 6

a) In Fig. 9 △ **ABC** is isosceles. If ∠**C = 30°** find the other angles.

   Then ∠**B** must be equal to **30°**. ∠**A** + ∠**B** +∠**C**= 180; but ∠**B** + ∠**C** = **60°**. ∠**A** = 180 - 60 = **120°**

b) In Fig. 9, with △ **ABC**, if ∠**A = 100°**, find the other angles.

   The sum of the other angles must be equal to **180 - 100 = 80°**. Each angle must be equal to **40** degrees since ∠**B** = ∠**C** = **40°**

c) Refer to fig **6**. If *a* = **130°** find *b, c, d, e, f, g* and *h*.
   a + *b* = **180°**; *b* = 180 - 130 = **50°**
   a = *d* (vertically opposite angles)
   a = *e* (corresponding angles)
   d = *h* (corresponding angles)
   a = *d* = *e* = *h* = **130** degrees
   b = *f* (corresponding angles)
   b = *c* (vertically opposite angles)
   c = *g* (corresponding angles)
   b = *f* = *c* = *g* = **50°**

d) In Fig. 9, if $\angle MLS = 120°$, and $\angle M = 50°$, find $\angle N$.

$$\angle MLS = \angle M + \angle N$$
$$\angle N = \angle MLS - \angle M$$
$$\angle N = 120° - 50°$$
$$\angle N = 70°$$

Example7

a) Find the measure of a, b c, e and f in Fig.10.

$f = 60○$ (alternate angles Z formation)
$f = e = 60○$ (base angles of an isosceles triangle)
$b = 180° - (e + f)$
$b = 180° - 120°$
$b = 60○$
$c = b + f$ (Exterior angle = sum of the interior opposite angles)
$c = 60° + 60° = 120○$
$c = 120○$
$a + c = 180○$ (interior angles on the same side of the transversal $= 180○$; C formation)
$a + 120° = 180°$; $a = 180° - 120° = 60°$
$a = 60°$, b $= 60°$, $c = 120°$, $e = 60○$, $f = 60°$

Fig 10

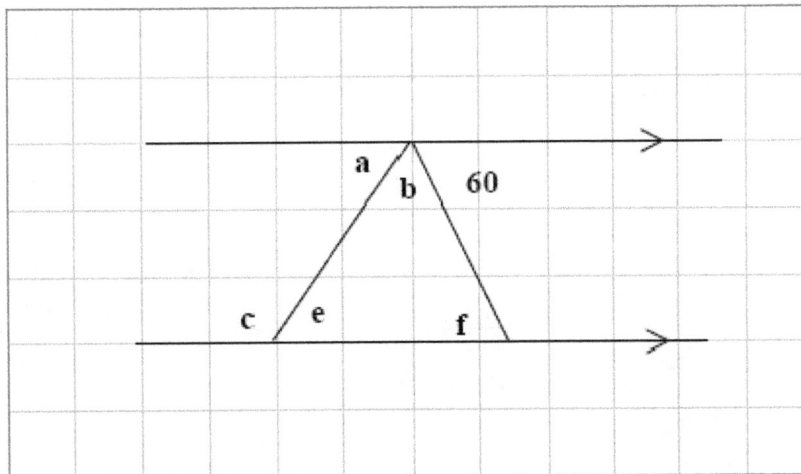

Practice

1.

    a) Find the values of *a*, *b* and *c*.

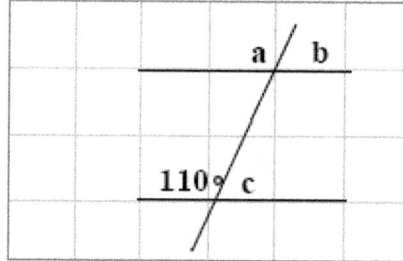

    b) Identify the angles corresponding to the **Z** and **C** formations in Fig. 1

Fig 1

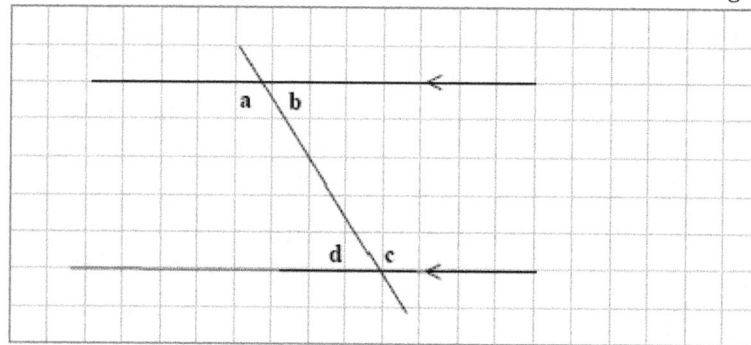

    c) Find the missing values in Fig. 2 **A**, **B**, **C** and **D**.

Fig 2

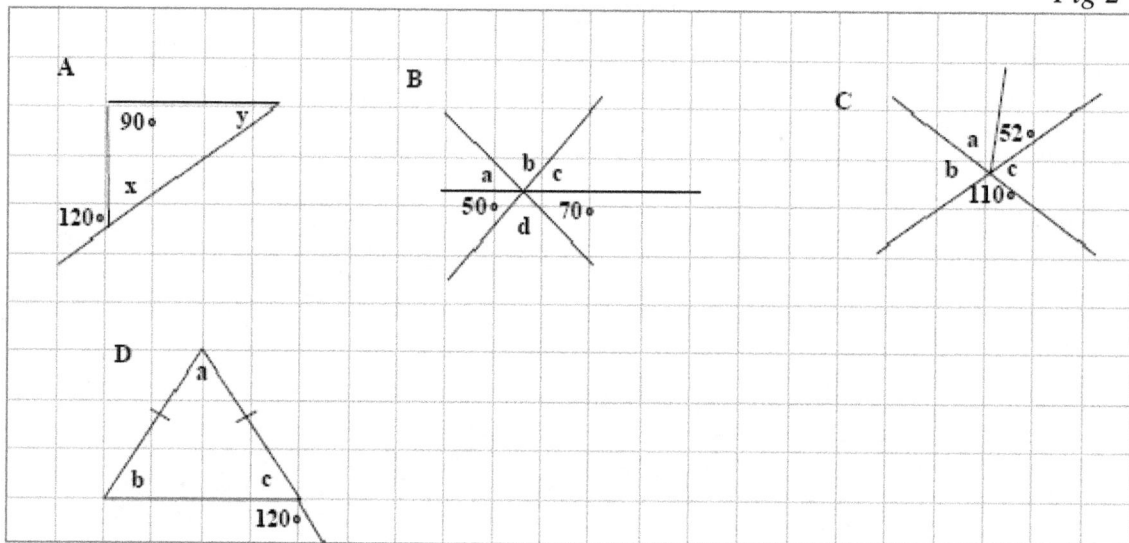

    d) **ABC** is an isosceles triangle with **AB = AC**. If angle **A = 50°**. Find angles **B** and **C**.

2. Find the lettered angles in Fig. 3 marked **A** and **B**.

Fig 3

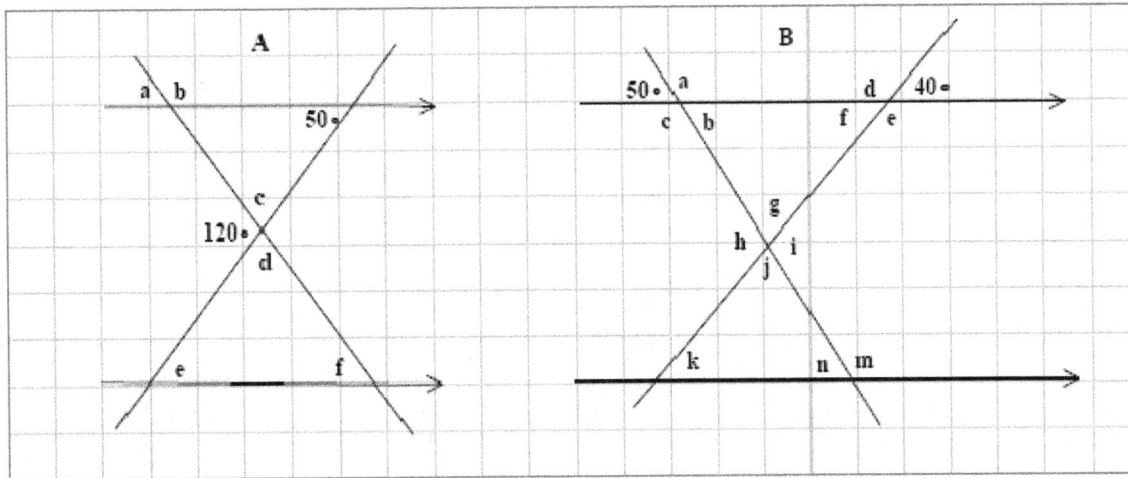

3. Find angles **B** and **D** of the kite in Fig. 4.

Fig 4

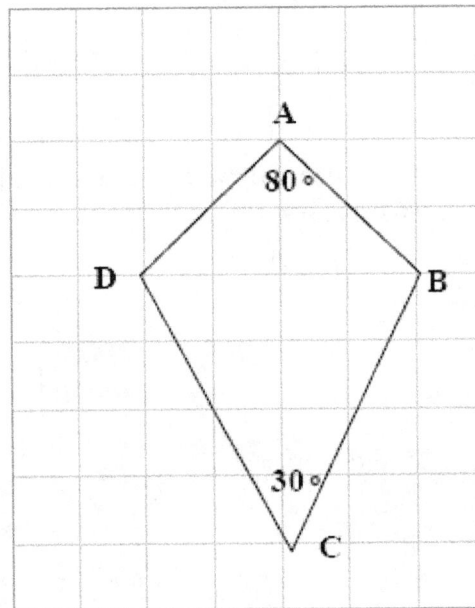

CHAPTER 7

BEDMAS 1

When doing arithmetic calculations using brackets, exponents, division, multiplication, addition and subtraction, a certain order of operations has to be used, referred to as BEDMAS.

B = Brackets
E = Exponents
D = Division
M = Multiplication
A = Addition
S = Subtraction

Calculations are done with Brackets first, next Exponents, then Division in that order and all the way down to subtraction.

Example 1

4 + 3 x 6	According to the rules of BEDMAS, multiplication comes before addition.
= 4 + 18	(*First multiply **6 x 3** and then add **4***)
= 22	

Example 2

(4 - 3 + 6) + 6 x 2	(Brackets are done first, then the multiplication and finally the addition)
= 7 + 12	Explanation **(4 - 3 +6) = 7; 6 x 2 = 12**
= 19	

Example 3

$$\frac{(5 \times 5 + 3) - (6 + 4 \times 3)}{6 + 7}$$

$$= \frac{(25 + 3) - (6 + 12)}{13} \qquad 5 \times 5 = 25 \text{ and then add } 3$$

$$= \frac{28 - 18}{13}$$

$$= \frac{10}{13}$$

The horizontal line means that everything above the line is divided by the denominator, which is **13** in this case.

The sign outside a bracket is multiplied by the sign inside a bracket.

Example 4

- (- 3) = + 3   (minus multiplied by minus equals plus)
- (7) = - 7   (minus multiplied by plus equals minus)

Example 5

(-4 + 3)5 + 8 (Brackets are done first; -4 +3 = -1)
= - 1 x 5 + 8
= - 5 + 8
= 3

Brackets means multiply. The number or variable outside the bracket is multiplied by whatever is inside the bracket as shown next.

  6 (4 + 7)
= 6(11)
= 66

Example 6

  4 x 2^3- 6(4 ÷ - 2 + 6) + 2   *(Exponent is done first 2 x 2 x 2 = 8, then multiply by 4)*
= 4 x 8 - 6(- 2 + 6) + 2   *(In the bracket, the division is done first 4 ÷ - 2 = - 2)*
= 32 – 6(4) + 2
= 32 – 24 + 2
= 34 - 24
= 10

Example 7

= - {4 + 6 ÷ (- 3)+ 8} – 4 x 5(6 - 2 x 6)
= - {4 - 2 + 8} - 20(6 - 12)
= - {10} - 20(- 6)
= - 10 + 120
= 110

When there is more than one set of brackets, start with the innermost brackets first.

Example 8

  {4 + 6 x 5 - (8 x 8 ÷ - 4 + 6) 4 + 9 x 4}{6 x 4 + 8}
= {4 + 30 -(8 x - 2 + 6) 4 + 36}{24 + 8}
= {34 - (-16 + 6)4 + 36}{32}
= {34 - (- 10) 4 + 36}{32}
= {34 + 10 x 4 + 36}{32}   *(Notice the inner brackets have disappeared)*
= {34 + 40 + 36}{32}
= {110}{32}
= 3520

Example 9

$$\left(\frac{11}{2} + 6 \right) - \left(\frac{7}{2} + 8 \right)$$

$$\frac{11 + 12}{2} - \left(\frac{7 + 16}{2} \right)$$

$$\frac{23}{2} - \frac{23}{2} = 0 \qquad \left(\frac{23}{2} \text{ and } \frac{-23}{2} = 0; \right. \text{ positive and negative numbers of the same value are}$$

equal to zero). Refer to Chapter 1 on Integers.

Practice 1

a) $(4 \div - 2 + 8 \div 2) - (7 + 3 \text{ x } - 4)$

b) $\left(\frac{11}{4} + 6 \right) - \left(\frac{7}{4} + 8 \right)$

c) $8 \div - 4 + 6(8 \text{ x } 2 - 2 + 4) + 6$

d) $-9 - (6 \text{ x } - 3 + 4 \text{ x } - 2 + 7)$

e) $-(8 \div - 2 - 7 \text{ x } 3 - 6 \div - 2)(- 4 \text{ x } 3 + 21 \div - 7)$

f) $3 \text{ x } 4 + 8 \text{ x } 3 - 5 \text{ x } 6 + 8 \text{ x } 4 - 8$

g) $-(4 \text{ x } 2 - 4 \div - 2 + 6 \text{ x } 3)(- 1)$

h) $-(- 6 \text{ x } 2 + 7 \text{ x } - 2) - 8 \text{ x } 3$

i) $5 \text{ x } [4 - (- 7)] - 2$

j) $6 + (8 \div - 2) \div 2$

k) $- 3(6 \div - 3 \text{ x } 3^2)\{(3 \text{ x } 2^3 - 7(- 6 + 4 \text{ x } 6^2)\}$

l) $- 6 \{ 2 \text{ x } 4^3 + 6(4 \div 2 + 2^2)(6 - 3 \text{ x } 2^2 + 7 \text{ x } 2^2) - 8 \}$

2. Determine the missing value in the following (x means multiply not the variable *x*)
    a) $- 36 \text{ x } \lozenge = 144$ b) $27 \text{ x } \lozenge = 135$ c) $\lozenge \text{ x } 15 = 90$ d) $\lozenge \text{ x } 5 = - 165$

3. Find the missing values.
    a) $-33 \div \lozenge = 3$ b) $- 144 \div \lozenge = 24$ c) $\lozenge \div 7 = - 8$ d) $\lozenge \div 19 = - 5$

CHAPTER 8

TRANSFORMATIONS

Translations on a coordinate system

$\triangle$**ABC** is translated **9** units to the right and **2** units down. The image of the translation is $\triangle$**A' B' C'** shown in Fig. 1

Fig 1

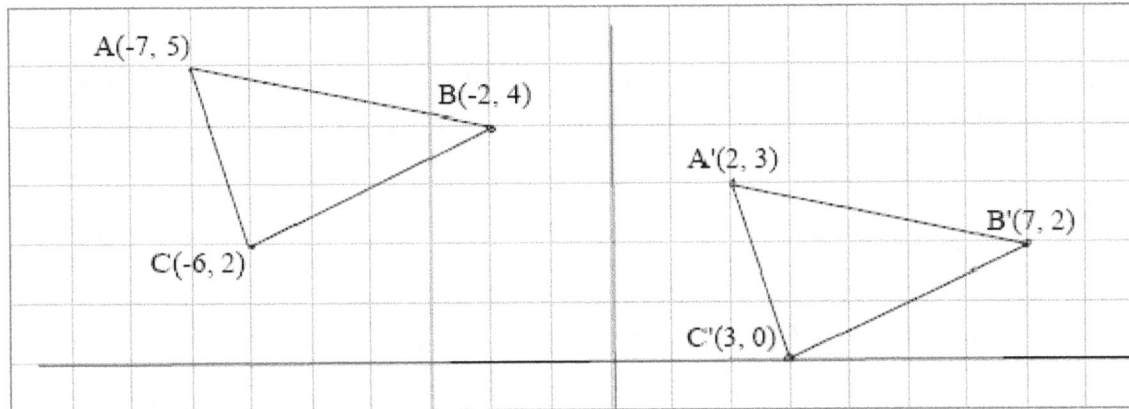

Since $\triangle$**ABC** is translated **9** units to the right, **9** is added to the **x** coordinates of $\triangle$**ABC** to get the image $\triangle$**A'B'C'**.
Translations to the right or left affect only the *x* coordinate

$\triangle$**ABC** is also translated down **2** units, so **2** is subtracted from the **y** coordinates of $\triangle$**ABC** to get the image $\triangle$**A'B'C'**
Translations up or down affect only the *y* coordinate

Translations up, add to the y coordinate and translations down, subtract from the y coordinate to get the image.

Translations to the right, add to the **x** coordinate and translations to the left, subtract from the **x** coordinate to get the image.

Example 1

Triangle $\triangle$**ABC** has coordinates **A(7, 2), B(4, 5), C(6, 2).**

Find the coordinates of the image **A', B', C',** if $\triangle$**ABC** is translated **6** units to the left.
Since$\triangle$ **ABC** is translated **6** units to the left, subtract **6** from the **x** coordinates of $\triangle$**ABC.** The **y** coordinates are not affected.
The coordinates of the image are **A'(1, 2), B'(-2, 5), C'(0, 2)**

Given the image **A'(1, 2), B'(-2, 5), C'(0, 2)** which is the result of a translations of **6** units to the left, find the coordinates of △ABC.

To get the coordinates of △ABC, we must add **6** to the **x** coordinates of △A'B'C'. Adding **6** to the **x** coordinates of △A'B'C' results in A(7,2) B(4,5) and C(6, 2) which are the original coordinates of △ABC.

If the image is the result of the original being translated **7** units to the right, then the original coordinates are obtained by moving the image **7** units to the left (the opposite).

Example 2

The coordinate of △A'B'C' are A'(4, -3) B'(-4, 6) , C'(2, -4) which is the result of the △ABC being translated **4** units to the right and **6** units down.
Find the coordinates of △ABC.

Subtract **4** units from the **x** coordinates and add 6 units to the **y** coordinates of the image △A'B'C' to get△ ABC.

The coordinates of△ ABC are A(0, 3) B(-8, 12), C(-2, 2)

Example 3

Fig 2

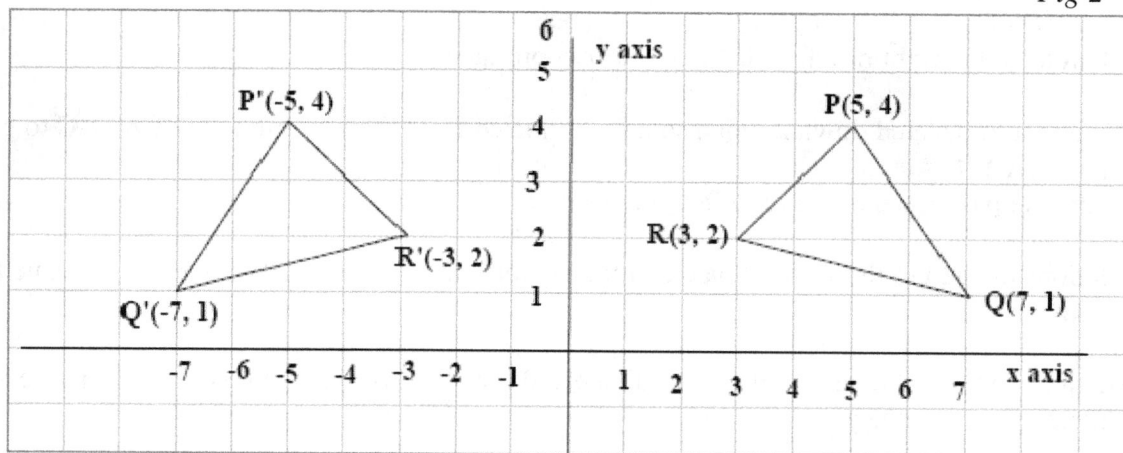

In Fig 2 △**PQR** is reflected on the **y** axis. In the reflected image △**P'Q'R'** , the sign on the **x** coordinate has changed; the **y** coordinate stays the same.

Example 4

Fig 3

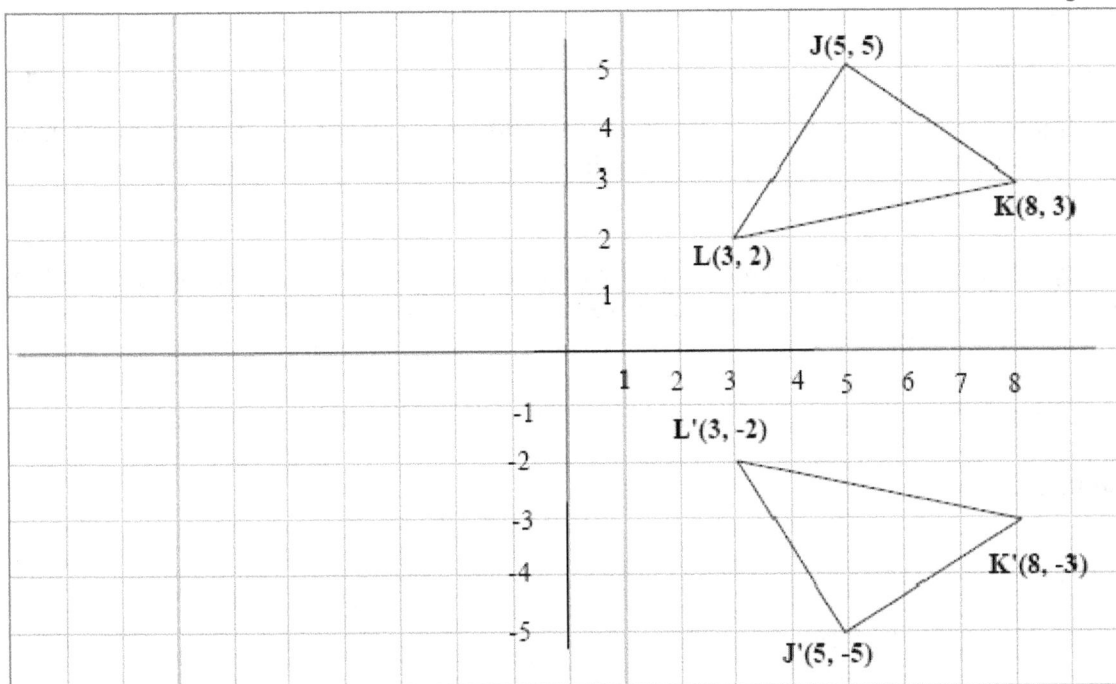

In Fig 3, △**JKL** is reflected on the **x** axis. In the reflected image △**J' K' L'**, the sign on the **y** coordinate changes while the **x** coordinate remains the same.

Example 5

Fig 4

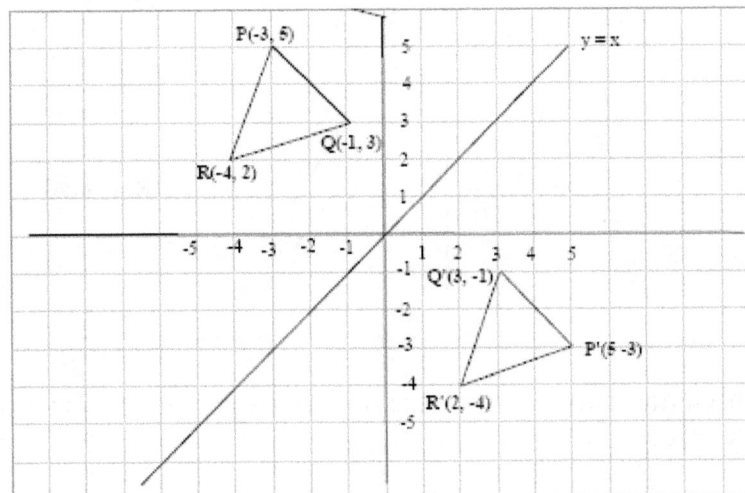

In Fig 4 △**PQR** is reflected on a line **y = x** (**45** degrees with reference to the **x** axis). x is the same as y . (If **x** is **4** then **y** is **4** on the line.)

The coordinates of the reflected image are reversed.
P(-3, 5) , P'(5, -3) ; Q(-1, 3) , Q;(3, -1) ; R(-4, 2) , R'(2, -4)

Note

If any figure has been translated **4** units to the left and then **2** units up, to get the image, then the image coordinates would have to moved **4** units to the right and **2** units down to get the coordinates of the original figure.

If any figure has been translated **6** units to the right and then reflected on the **x** axis to get the image, then the image coordinates would have to be first reflected on the **x** axis and then translated **6** units to the left to get the coordinates of the original figure.

The image transformations are in the reverse order to the original transformations.

Example 6

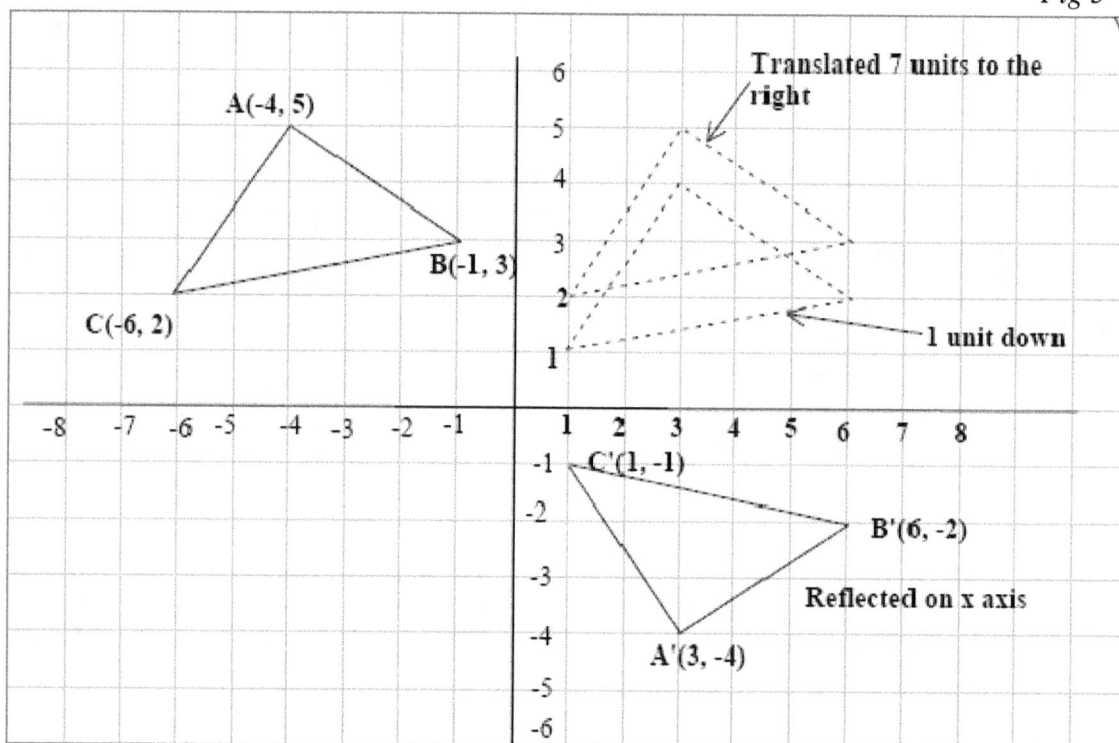

Fig 5

In Fig 5, △**ABC** has been translated **7** units to the right and **1** unit down. The resulting image is then reflected on the **x** axis.

Given the coordinates of the image, how can the coordinates of **ABC** be calculated?
First, reflect the coordinates of the image **A'B'C'** on the **x** axis, then translate the coordinates up **1** unit and to the left **7** units.

Reflect **A'(3, -4)** on the **x** axis; the coordinates become **(3, 4)**. Move up **1** unit; the coordinates become (**3, 5**). Translate **7** units to the left. (**-7 + 3 = -4**); the coordinates now become **A(-4, 5)** which matches the coordinates of **A** in Fig 5.

Reflect **B'(6, -2)** on the *x* axis; the coordinates become (**6, 2**). Move **1** unit up; the coordinates become (**6, 3**). Move **7** units to the left; the coordinates become **B(-1, 3)** which matches **B** in the figure.

Reflect **C'(1, -1)** on the x axis; the coordinates become (**1, 1**). Move up **1** unit; the coordinates become (**1, 2**). Move **7** units to the left; the coordinates become **C(-6, 2)** which corresponds with the coordinates in Fig 5

Example 7

Fig 6

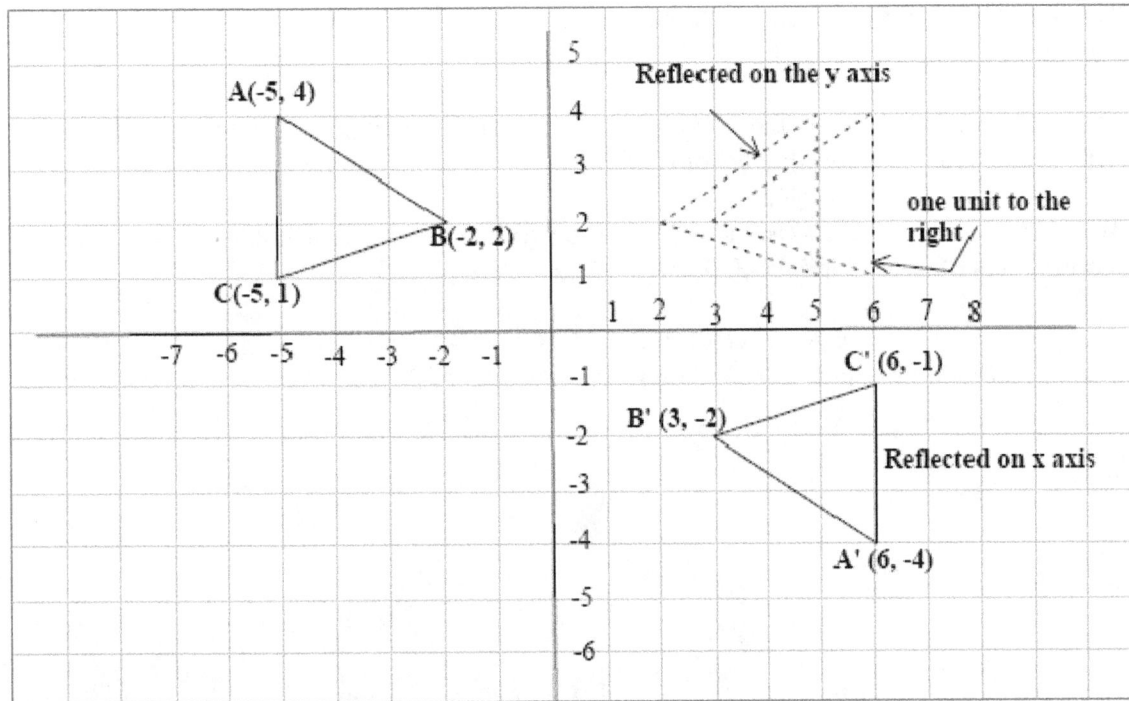

In Fig 6 Δ **ABC** has been reflected on the **y** axis, then moved one unit to the right and finally reflected on the **x** axis.

Calculate the coordinates of **A' B' C'** based on the transformations and check to see if they match the coordinates **A' B' C'**.

Practice

1. **A(3, 4)** , **B(-4, 3)** . **C(-6, 2)** has been translated **4** units to the left and **6** units down. Find the coordinates of the image Δ **A' B' C'**?

2. Δ **XYZ** has been translated **5** units to the left and **4** units down. If the reflected image Δ **X'Y' Z'** has coordinates, **X'(5. 3)**, **Y'(-6, 4)**, **Z'(2, -4)**, find the coordinates of **ΔXYZ**

3. Δ**KLM** with coordinates **K(4, -5)**, **L(2 -4)**, **M(-4, 3)** is reflected on the *x* axis. Find the coordinates of its image Δ **K'L'M'**

4. The coordinates of Δ**PQR** are P(3, -5), Q(-3, -4), R(2 -6) are reflected on the *y* axis. Find the coordinates of ΔP'Q'R' (The image of PQR)

5. The coordinates of Δ**JKL** are **J(-4, 3)**, **K(5, -6)**, **L(-1, 6)**. If **JKL** is reflected on the *y* axis , then translated **5** units up and **4** units to the left, calculate the coordinates of the image **J'K'L'**.

6. The coordinates of the image of Δ**ABC** are A'(2, -4), B'(6. 3), C'(-2, 5). Δ**ABC** has been reflected on the *x* axis and then translated **6** units to the right and **5** units up. Calculate the coordinates of Δ**ABC**.

7. Δ**WXY** has been reflected on the line *y = x*. If the coordinates of Δ**WXY** are **W(2, 5)**, **X(-7, 4)**, z(6 -4), what are the coordinates of the image?

8. Δ**ABC** is reflected on the x axis and then on the *y* axis. Find the coordinates of the image if the coordinates of Δ**ABC** are **A(-3, 5)**, **B(6, -4)**, **C(-7, 4)**?

9. The coordinates of the image of Δ**JKL** are **J'(4, -6)**, **K'(-3, 7)**, **L'(5, -4)**. If **J'K'L'** is the result of a reflection on the *x* axis and then on the *y* axis of Δ **JKL**, what are the coordinates of **JKL**?

CHAPTER 9

PROBABILITY

Probability is the possibility of a result of some action. When a coin is tossed, what is the probability of getting heads? There are only two possible outcomes either a head or a tail. Suppose a coin was tossed and someone bet on heads, then heads is the favourable outcome.

The probability of getting heads = $\dfrac{\textbf{Favorable outcome}}{\textbf{total outcomes}} = \dfrac{1}{2}$

The total outcomes are **2** as the coin can only show heads or tails.

If a die with six sides is rolled, the probability of getting a **2** is $\dfrac{1}{6}$

Experimental and theoretical probability

If a coin is tossed, in theory, heads should come up **50%** of the time and tails **50%** of the time, but in practice this would rarely be the case. Toss a coin **32** times and register the heads and tails with a tally chart. The experimental result would not necessarily be exactly **16** heads and **16** tails.

The experimental and theoretical probability is not always the same.

Below is the result of **32** tosses of a coin. There were **15** heads and **17** tails.

Outcome	Tally	Frequency
Heads	ℍℍ ℍℍ ℍℍ	15
Tails	ℍℍ ℍℍ ℍℍ ‖	17

The tally chart was filled in for each toss of the coin and then the frequency column was filled in to match the tally.

Example

If a coin is tossed and a die with six sides is rolled, what are the possible outcomes?
This can be shown with a tree diagram, on the next page.

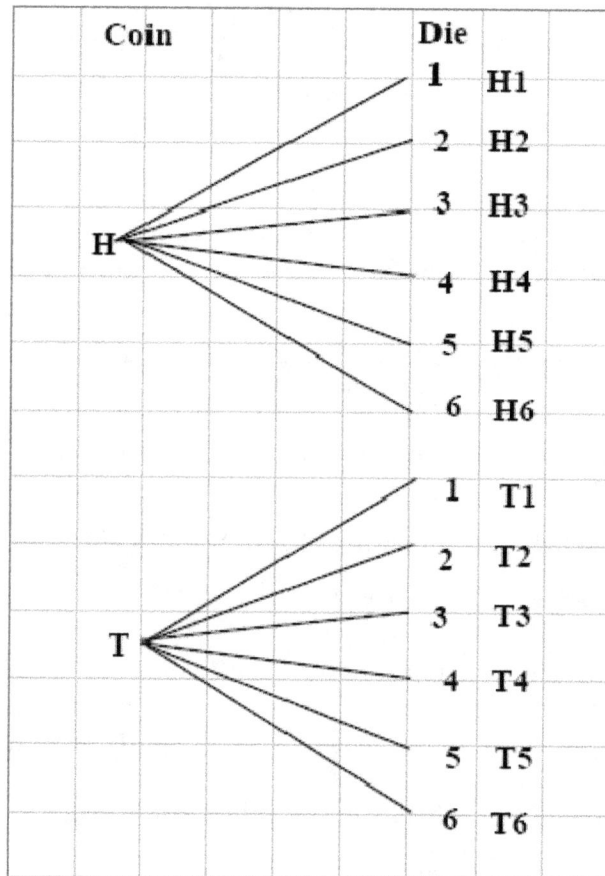

There are **12** outcomes.

What is the probability of a head and a three showing up? From the tree diagram the probability is one out of **12** or $\dfrac{1}{12}$

Example

If two six sided dice are rolled, the result of two events can also be shown in a table. In this case the table is showing the sum of the two rolls.

From the table, the probability of getting a sum of **5** is $\dfrac{4}{36} = \dfrac{1}{9}$

The probability of getting a sum greater than **8** is $\dfrac{10}{36} = \dfrac{5}{18}$

	Die 1					
	1	2	3	4	5	6
1	2	3	4	5	6	7
2	3	4	5	6	7	8
Die 2 3	4	5	6	7	8	9
4	5	6	7	8	9	10
5	6	7	8	9	10	11
6	7	8	9	10	11	12

Example

In a family with three children, what is the probability of having:

 a) Three boys
 b) Two girls and a boy
 c) Two boys and girl
 d) Three girls

See the tree diagram on the next page.

a) $\dfrac{1}{8}$

b) $\dfrac{3}{8}$

c) $\dfrac{3}{8}$

d) $\dfrac{1}{8}$

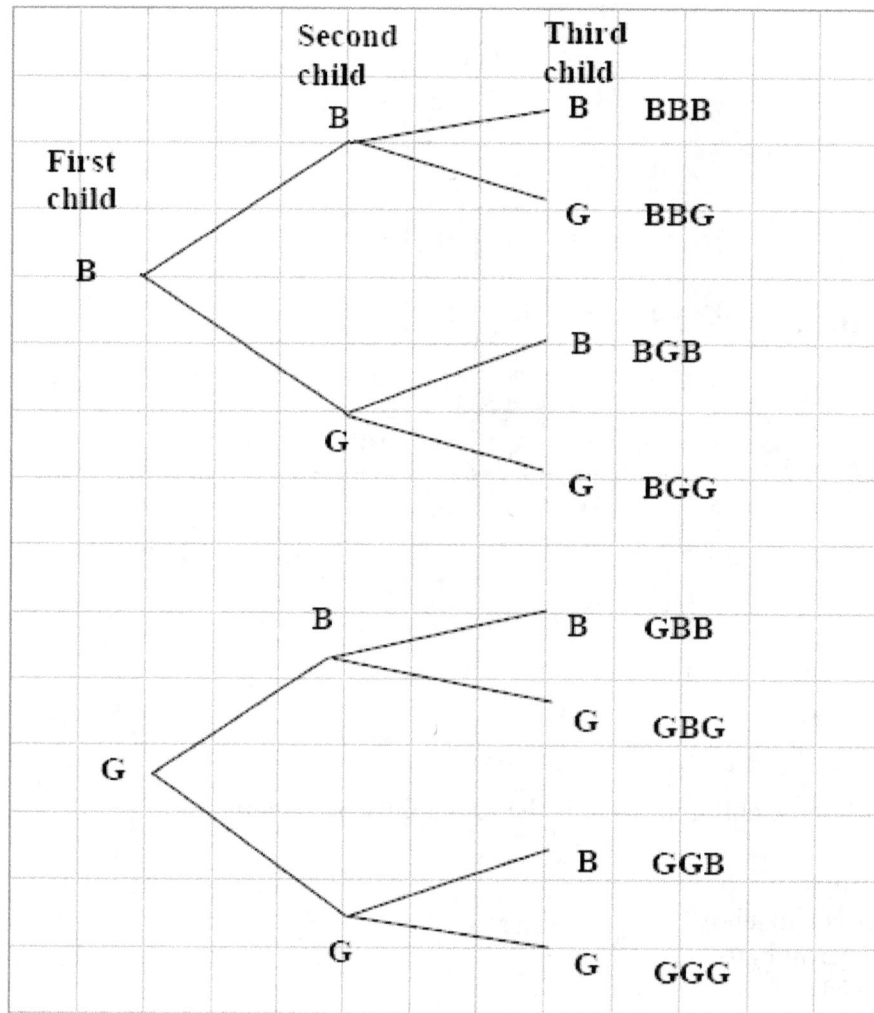

	Second child	Third child	
		B	BBB
	B		
		G	BBG
First child			
B		B	BGB
	G		
		G	BGG
		B	GBB
	B		
		G	GBG
G			
		B	GGB
	G		
		G	GGG

The total outcomes with a combination of events are obtained by multiplying the outcomes of each event by the outcomes of the other events. In the above example, the outcome for each event is **2**. Since there are three events, the total outcome of all three events is equal to **2 x 2 x 2 = 8**.

If a coin is tossed and a six sided die is rolled, then the total outcomes will be **2 x 6 = 12**.

Practice

1. If a **6** sided die is rolled and a spinner with four colors, red, blue, green and yellow is spun, what is the probability of each event?

   Draw a tree diagram

   a) P(**2** and Blue) (Probability of **2** and Blue)
   b) P(number > **4** and Red)

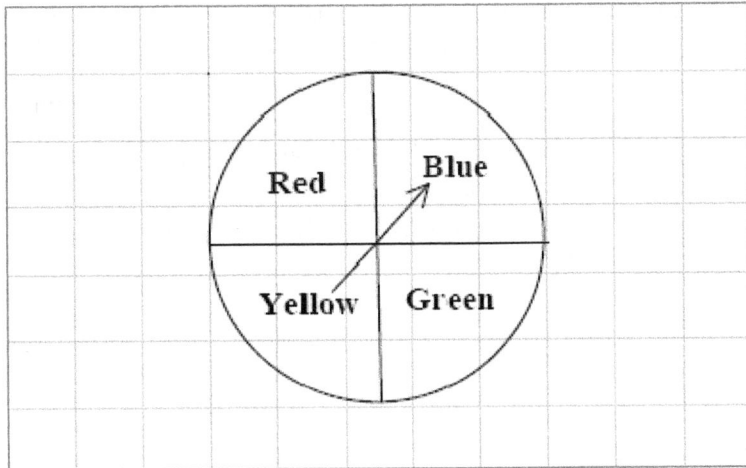

2. Karim has to spin the spinner twice to see how much he will win in a contest.

   Draw a table to show all possible sums of two spins.

   a) What is the probability that he will win **$900** after two spins?
   b) What is the probability that he will win more than **$1300** after two spins?

   Give the answer in fractions and percents.

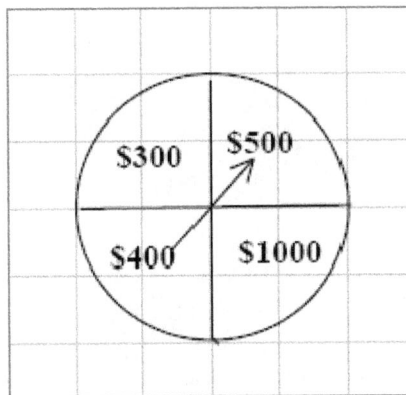

3. What is the probability of each event?

   a) Drawing a queen from a deck of cards.
   b) Drawing black **10** from the deck.

4. If a couple decide to have four children, what is the probability of having

   a) Three girls and a boy?
   b) Two girls and two boys?
   c) All four girls?

5. Jan has three pairs of pants: a red pair, a black pair and a brown pair. She has four hats: one green, one black, one brown, and one grey. She also has three jackets: one white, one pink and one yellow.

   a) Use a tree diagram to show all possible combinations of outfits Jan can wear.
   b) What is the probability that her outfit will include something black?
   c) What is the probability that her outfit will include something yellow or brown?
   d) What is the probability that her outfit will include yellow and brown?

CHAPTER 10

REVIEW OF ESSENTIAL SKILLS FOR GRADE 9

Knowledge of integers, fractions percentages rates ratios and equations are necessary skills for Grade 9.

Refer to previous chapters on these topics covered in Grade 8 in this book.

More on fractions

When mixed numbers are used, always convert to improper fractions before performing fraction operations.

Example

$$-3\frac{2}{3} x\, 4\frac{1}{4} = \frac{-11}{3} x\, \frac{17}{4} = -\frac{187}{12} = 15\frac{7}{12}$$

Example

$$4\frac{3}{4} \div 7\frac{5}{8} = \frac{19}{4} \div \frac{61}{8} = \frac{19}{4} x\frac{8}{61} = \frac{152}{244} = \frac{38}{61}$$

Another way of showing fraction division

$$\frac{\dfrac{3}{4}}{\dfrac{1}{2}}$$

$$= \frac{3}{4} \div \frac{1}{2}$$

$$= \frac{3}{4} x\, \frac{2}{1} = \frac{3}{2}$$

Example

$$1 + \cfrac{1}{1+\cfrac{1}{1+\frac{1}{2}}} \qquad (Start\ from\ the\ bottom\ and\ work\ your\ way\ up)$$

$$= 1 + \cfrac{1}{1+\cfrac{1}{\frac{2+1}{2}}}$$

$$= 1 + \cfrac{1}{1 + \cfrac{1}{\frac{3}{2}}}$$

$$= 1 + \cfrac{1}{1 + \cfrac{2}{3}} \qquad \{1 \div \frac{3}{2} = \frac{2}{3}\} \quad 1 + \cfrac{1}{1 + \frac{2}{3}}$$

$$= 1 + \cfrac{1}{\cfrac{5}{3}}$$

$$= 1 + \cfrac{3}{5}$$

$$= 1\frac{3}{5}$$

Note

$$= 1 + \frac{3}{5} \text{ is the same as } 1\frac{3}{5}$$

Percent

Example

The value of a house increased from **$250,000** to **$275,000** in one year. What was the percent increase?

$$\text{The percent increase} = \frac{\textbf{The increase}}{\textbf{original value}} \times \textbf{100}$$

$$= \frac{25,000}{250,000} \times 100 = 10\%$$

If the value of a house increased **10**% and the original price was **$300,000**, what is the new price of the house?

$$\frac{10}{100} x \ 300000 = 30,000$$

The new price of the house is **300,000 + 30, 000 = $330,000.**

Recall from the previous chapters on percent that the same answer can be calculated more quickly by multiplying the price of the house by **1.10.**

The new price is **300,000** x **1.1 = $330,000**

Note

1 + 0.1 is the same as **1.1** as shown below

$$\begin{aligned} \text{add} \\ 1.00 \\ \underline{+0.1} \\ =1.1 \end{aligned}$$

If there is an increase in percent, add the increase to **1**. If there is a decrease in percent, subtract the decrease from **1.**

If a car depreciates (value decreases) by **8%** a year, then multiply the price of the car by **1 – 0.08 = 0.92** to get the discounted value.

Example

The price of a new car is **$23,000**. What is the price one year from now if the car depreciates by **11%** a year?

The depreciated price is **23000** x **0.89** = **$20470**

The inverse of an operation or working backwards

If **100** is multiplied by **5** the result is **500**.
Reverse the operation If **500** is divided by **5** the result is the same:- **100**.

Example

$$130 \times 15 = 1950$$
$$\frac{1950}{15} = 130$$

Multiplication and division are opposite operations as shown above

Example

A gold coin appreciates (increases in value) by **7%** a year. If the price of the coin today is **$450**, what was the price last year?

To get the price of the coin one year back, divide the today's price by **1.07.**
Working backwards, last year's price = $\dfrac{450}{1.07}$ = **$ 420.56**
Multiply this year's price by **1.07** to get next year's price.
Divide next year's price by **1.07** to get this year's price. (Of course we don't know next year's price but this example shows you how the inverse operation works)

Another method to solve this problem is to write it as an equation (refer to chapter **5**)

Last year's price x **1.07** = **450** (This year's price)

$$\text{Last year's price} = \frac{450}{1.07} = \$420.56$$

The same problem can be solved by using ratios.

$$\frac{\text{Today's price}}{\text{Last year's price}} = \frac{\text{Today's price}}{\text{Last year's price}}$$

$$\frac{1.07}{1.00} = \frac{450}{\text{Last year's price}}$$

*(If last year's price is **1.00**, then today's price equals **1.07**)*

$$\text{Last year's price} = \frac{450}{1.07} x\ 1.00 = \$420.56$$

If last year's price of **$420.56** was substituted in the same ratio, then this year's price of **$450** would be obtained.

Another example on percent

25% of what is **70**   (Meaning, when a number is multiplied by **0.25** the answer is **70**)

Working backwards; <u>to find the number</u>, divide **70** by **0.25**

$$\frac{70}{0.25} = 280$$

The answer is **280**

Or writing this as an equation.

Let the number be x

$$0.25\ x = 70$$

$$x = \frac{70}{0.25} = 280$$

Practice

1. Evaluate

    a) $\frac{1}{2}\left(\frac{1}{3} - 2\frac{1}{5}\right)$

    b) $-\frac{1}{2}\left(-3\frac{1}{4}\right) - \frac{1}{3}$

    c) $-\frac{5}{6} x\ 2\frac{1}{4} - \left(-1\frac{3}{4}\right)$

    d) $\frac{21 - 2.4\ x3.6}{15 + 4 \div -2}$

e) $2 + \dfrac{1}{3 + \dfrac{2}{3}}$

f) $2 + \dfrac{1}{1 + \dfrac{2}{3}}$

2. (refer to ratios rates and percent)
   a) A car can do one trip on $\dfrac{1}{8}$ of a tank of gas. If the tank is $\dfrac{3}{4}$ full, how many trips can the car make?
   b) A blade of grass grew from $5\dfrac{1}{8}$ **cm to** $5\dfrac{3}{8}$ **cm**. By what percent did the blade of grass grow? Express as a fraction and then multiply by **100**.
   c) The purchase price of a coat was $89. If the sales tax is **15%** and the coat was discounted by **25%**, what was the price of the coat before the discount?
   d) What is **21%** of **125**?
   e) What percent of **80** is **15**?
   f) **35%** of what number is **120**?

Algebraic expressions.

Examples

The wind blows at w km/h. Write an expression if
   a) The wind speed increases by **20** km/h.
   b) The wind speed is doubled
   c) The wind speed decreases by **30** km/h
   d) The wind speed is halved

Solution
   a) w +**20**
   b) **2**w
   c) w – **30**
   d) $\dfrac{w}{2}$

Practice

Write algebraic expressions for the following.

3. I have $3x + 5$ apples. Sam has **5** more than me. Write an algebraic expression for the number of Sam's apples.
4. Mary bought $(x+5)$ apples. She paid **0.25**c each. Write an expression for the cost of the apples.
5. A number increased by **6**.
6. Six times a number divided by **2**.

7. A number multiplied by itself.
8. **4** decreased by a number doubled.
9. A number increased by itself.
10. A number subtracted from **6**.
11. **5** subtracted from a number.
12. **4** subtracted from a number divided by **2**.

Equations

Example

Given the formula for the volume of a sphere: $V = \frac{4}{3}\Pi r^3$; express r in terms of V.

$$\frac{4}{3}\Pi r^3 = V$$

$$\Pi r^3 = V\left(\frac{3}{4}\right)$$

$$r^3 = \frac{V}{\Pi}\left(\frac{3}{4}\right)$$

$$r^3 = \frac{3V}{4\Pi}$$

Taking the cube root of both sides.

$$\sqrt[3]{r^3} = \sqrt[3]{\frac{3V}{4\Pi}}$$

$$r = \sqrt[3]{\frac{3V}{4\Pi}}$$

CHAPTER 11

EXPONENTS

Exponents were discussed briefly in Chapter 3 on fractions decimals and percents, but in this Chapter it will be discussed fully.

4 x 4 x 4 can be shown as 4^3 .

4^3 is the exponential form. **4** is the base and **3** is the power.

4 is raised to the power of **3** is another way of expressing 4^3

Laws of exponents

1. $a^2 \times a^3 = a^5$ ($a^2 = a \times a$; $a^3 = a \times a \times a$; $a^2 \times a^3 = a \times a \times a \times a \times a = a^5$). When exponents of the same base are multiplied, add the exponents algebraically.
$a^3 \times a^{-2} = a^1 = a$

2. $\dfrac{a^5}{a^3} = a^{5-3} = a^2$ $\left\{ \dfrac{a \times a x \, a \times a \times a}{a \times a \times a} = a^2 \right\}$

When exponents of the same base are divided, subtract the exponent in the denominator from the exponent in the numerator.

x^3 also reads as x raised to the power of **3**

3. $\left(a^3\right)^4 = a^{12}$ Multiply the exponent in the bracket by the exponent outside the bracket.
$\left(a^3\right)^4 = a^3 \times a^3 \times a^3 \times a^3 = a^{12}$

4. $\left(\dfrac{a^3}{b^4}\right)^2 = \dfrac{a^6}{b^8}$ The numerator and denominator have different bases. The exponent of each base is multiplied by the exponent outside the bracket.

5. $(2\,ab)^4 = \left(2^4\right)\left(a^4\right)\left(b^4\right)$
$= 16\,a^4 b^4$

6. $\left(\dfrac{a}{b}\right)^4 = \dfrac{a^4}{b^4}$

Negative exponents are the reciprocals of positive exponents

The reciprocal of a number is obtained by dividing <u>one</u> by the number.

The reciprocal of **4** is $\dfrac{1}{4}$

$$3^{-2} = \dfrac{1}{3^2} \quad \textit{(Negative exponents are reciprocals of positive exponents)}$$

Another way of looking at negative exponents is that the sign changes when the exponent is moved from the numerator to the denominator or from the denominator to the numerator.

$$\dfrac{1}{4^{-3}} = \dfrac{4^3}{1} = 4^3$$

Example 1

Express as a single power.

a) $6^6 \times 6^4$
b) $b^5 \times b^8$
c) $m^{-6} \times m^{-4}$

Solution

a) $6^6 \times 6^4 = 6^{10}$
b) $b^5 \times b^8 = b^{13}$
c) $m^{-6} \times m^{-4} = m^{-10} = \dfrac{1}{m^{10}}$

$$\textit{(always express as a positive power)}$$

Example 2

Express as a single power.

a) $\dfrac{7^4}{7^{-5}}$
b) $\dfrac{5^{-4}}{5^{-5}}$

Solution

a) $\dfrac{7^4}{7^{-5}} = 7^4 \times 7^5 = 7^9$
$\quad \textit{Explanation} \quad \dfrac{7^4}{7^{-5}} = \dfrac{7^4}{1} x \dfrac{1}{7^{-5}} = 7^4 \times 7^5$

b) $\dfrac{5^{-4}}{5^{-5}} = 5^{-4} \times 5^5 = 5$
$\quad \textit{(sign of the exponent changes when moving from the denominator}$

$$\textit{to the numerator)}$$

$$\text{or} \quad \dfrac{5^{-4}}{5^{-5}} = 5^{-4-(-5)} = 5^{-4+5} = 5$$

Example 3

Express as a positive power.

a) 4^{-5}
b) $\dfrac{1}{x^{-7}}$

a) $4^{-5} = \dfrac{1}{4^5}$ 　　　　 b) $\dfrac{1}{x^{-7}} = x^7$

Example 4

$$(4\,xy)^4 = 4^4 x^4 y^4 \quad \textit{(4xy means 4 multiplied by x multiplied by y)}$$
$$= 256x^4 y^4$$

The power of each number or variable in the <u>multiplication</u> in the bracket must be multiplied by the power outside the bracket.

<u>Note</u>　$(4 + x)^4$ is not equal to $4^4 + x^4$　(*This can be solved by Pascal's triangle which is grade 11 material*).

If the number and variable in the bracket are multiplied by each other, then the power of each number or variable in the bracket is multiplied by the power outside the bracket as in Example 4 above

Example 5

$$\left(\dfrac{3}{2}\right)^3 = \dfrac{3^3}{2^3} = \dfrac{27}{8} = 3\dfrac{3}{8}$$

Example 6

$$(-3)^{-3} = \dfrac{1}{(-3)^3} = \dfrac{1}{-27} \text{ same as } -\dfrac{1}{27} \text{ refer to the Chapter on fractions.}$$

Practice

1. Simplify

    a) $(a^5)(a^7)$　 b) $(a^3)(a^{-2})$　 c) $(5^6)(5^8)$　 d) $(3^4)(3^{-1})$　 e) $(m^5)(m^{-8})(m^4)$

2. Evaluate

    a) $\dfrac{2^6}{2^4}$　　　 b) $\dfrac{5^6}{5^4}$　　　 c) $\dfrac{6^7}{6}$　　　 d) $\dfrac{8^7}{8^5}$

3. Simplify

    a) $(3^6)^4$　　　 b) $(8^4)^3$　　　 c) $(10^4)^4$

4. Express as a single positive power.

a) $\dfrac{3^4}{3^3}$ 　　b) $\dfrac{4^5}{4^5}$ 　　c) $\dfrac{2^6}{2^8}$ 　　d) $\dfrac{3^{-5}}{3^7}$ 　　e) $\dfrac{3^{-3}}{3^{-3}}$

f) $2^4 \div 2^{-3}$ 　g) $10^6 \div 10^{-4}$ 　h) $\dfrac{c^{-7}}{c^{-5}}$ 　i) $\dfrac{s^{-3}}{s^{-6}}$ 　j) $\dfrac{\left(2^3\right)^{-2}}{\left(2^{-2}\right)^3}$

Simplify

5.　a) $\left(3\, a^3 b^2 x^3\right)^3$ 　　　　　　b) $\left(2\, x^2 b^3 a^5\right)^3$

Example 7

Simplify $\left(\dfrac{m^6}{n^5}\right)^{-4} = \dfrac{m^{-24}}{n^{-20}} = \dfrac{n^{20}}{m^{24}}$ (6x-4 = -24; 5x-4 = -20)

Example 8

Simplify $\left(\dfrac{m^5}{n^4}\right)\left(\dfrac{n^5}{m^3}\right)\left(\dfrac{m^6}{n^3}\right)$

$= \dfrac{m^5 \text{ x } m^{-3} \text{ x } m^6}{n^4 \text{ x } n^{-5} \text{ x } n^3}$

(The powers of *m* are **5**, **-3** and **6 = 8**; the powers of *n* are **4**, **-5** and **3 =2**)

$= \dfrac{m^8}{n^2}$

Some rules about numbers and variables to remember

Any number raised to the power of zero is equal to one.

Example 10

$\dfrac{6^2}{6^2} = 1$ 　　If the numerator and denominator are the same the result of the division is **1**

$\dfrac{6^2}{6^2} = 6^2 \text{ x } 6^{-2} = 6^0 = 1$ from the previous line.

$6^0 = 1$

The same applies to any variable such as $x^0 = 1.$ Any number or variable raised to the power of zero is equal to one.

Any number or variable divided by zero is undefined.

Example 11

$$\frac{c}{0} \text{ or } \frac{8}{0} \text{ is undefined.}$$

Any number or variable multiplied by zero is equal to zero.

Example 12

$$0\,(x) = 0$$
$$0\,(45) = 0$$

Zero divided by any number or variable is equal to zero.

Example 13

$$\frac{0}{8} \text{ or } \frac{0}{k} = 0$$

When a number or variable is written they are always in the numerator.

Example 14

$$x = \frac{x}{1} \;;\; 8 = \frac{8}{1}$$

Practice

6. Evaluate the following.

a) $\dfrac{c}{0}$ b) $(0)x$ c) $\dfrac{0}{6}$ d) $\dfrac{c^2}{c^2}$ e) c^0 f) $\dfrac{5\,xy}{5\,xy}$

Further examples on fraction multiplication and division

$(4)\left(\dfrac{1}{2}\right)$ is the same as $\dfrac{4}{2}$ $\left(\dfrac{4}{1} x \dfrac{1}{2} = \dfrac{4}{2}\right)$

$\dfrac{1}{3}y$ is the same as $\dfrac{y}{3}$

$\dfrac{4\,(3)}{4}$ is the same as $\dfrac{4}{4}(3) = 3$ $\left(\dfrac{4}{4} = 1\right)$

Similarly $\dfrac{4\,x}{4} = x$ (**4x** means **4** multiplied by x)

Examples

1. $\dfrac{4\,x}{x} = 4$

2. $\dfrac{4\,x^2}{x} = 4x$ ($\dfrac{x^2}{x} = x$ *refer to exponent law no **2***)

3. $\dfrac{8\,x^3}{4\,x} = 2x^2$

4. $\dfrac{12\,x^7 y^8}{3\,x^6 y^4} = 4xy^4$

5. $\dfrac{4\left(a^6\right)^{-4} b^6 c^7}{12\,a^6 b^7 c^9}$ *(moving b^6 and c^7 to the denominator)*

$= \dfrac{a^{-24}}{3\,a^6\,b^7 b^{-6} c^9 c^{-7}}$ *(moving a^{-24} to the denominator)*

$= \dfrac{1}{3\,a^6 a^{24} b c^2}$ *(There is a 1 in the numerator as $\dfrac{4}{12} = \dfrac{1}{3}$)*

$= \dfrac{1}{3\,a^{30} b c^2}$

6. $\dfrac{c^{-6}}{c^{-5}} = \dfrac{1}{\left(c^{-5}\right)\left(c^6\right)} = \dfrac{1}{c}$ *(This method saves a few steps; when an exponent is moved from the numerator to the denominator the sign on the power changes)*

Using the other method of subtracting the exponent of the denominator from the exponent of the numerator for ***Example 6.***

$\dfrac{c^{-6}}{c^{-5}} = c^{-6-(-5)} = c^{-6+5} = c^{-1} = \dfrac{1}{c}$

7. $\dfrac{a^{-4}}{a^{-5}} = \left(a^{-4}\right)\left(a^5\right) = a$ *(In this case moving the a^{-5} to the numerator results in a positive exponent +5 -4 =1)*

8. $\dfrac{b^{-7}}{b^{-5}}$ *Moving b^{-7} to the denominator would result in a positive exponent in the denominator.*

$= \dfrac{1}{\left(b^7\right)\left(b^{-5}\right)}$

$= \dfrac{1}{b^2}$

9. $-(4)^3 = -64$

10. $-(-4)^3 = -(-64) = 64$

11. $(-2)^3 = -8$ *If a negative base is raised to an odd number as in this case, then the answer will always be negative.*

12. $(-2)^2 = 4$ *If a negative base is raised to an even number the answer is always positive.*

13. $-(-2)^2 = -4$ *[$(-2)^2 = 4$ but the minus in front of the bracket makes the answer negative]*

14. $-4^{-3} = -\dfrac{1}{4^3} - \dfrac{1}{64}$

15. $\left(-3\,a^{-3}y^2\right)^{-3} = -\dfrac{1}{27} \times \dfrac{a^9}{y^6} = -\dfrac{a^9}{27\,y^6}$ $\{-3^{-3} = -\dfrac{1}{27}\}$

16. $\left(\dfrac{a^{-3}b^{-2}}{a^2 b^4}\right)^{-3} = \dfrac{a^9 b^6}{a^{-6} b^{-12}} = a^{15} b^{18}$

Practice

7.

   a) $-(-3)^3$ b) $(-4)^3$ c) $-(-5)^4$ d) $-(5)^3$ e) $-(-x)^3$

8. Simplify

   a) $\left(3\,x^4 y^{-3}\right)^2$ b) $\left(5\,y^3 b^{-4}\right)^{-2}$ c) $\left(4\,c^{-6}x^4\right)^{-3}$ d) $(-2)^3$ e) $\left(-2\,x^{-5}y^3\right)^{-3}$

                                                             refer to Example 15

   f) $\left(-2\,x^{-3}\right)^{-5}$ g) $\left(-3\,c^2\right)^4$ h) $\left(\dfrac{x^3 y^4}{x^{-2}y^{-3}}\right)^{-4}$ i) $\left(\dfrac{x^{-2}y^{-3}}{x^{-6}y^5}\right)^{-3}$

   j) $\left(\dfrac{5\,x^{-3}}{4\,y^{-3}}\right)^{-2}$ k) $\left(\dfrac{(2\,x)^3}{(3\,y)^2}\right)^{-3}$ l) $\left(\dfrac{x}{y}\right)^{-3}$ m) $\left(\dfrac{3}{4}\right)^{-1}$

          (Do the inner brackets first for k)

9. Evaluate

   a) $\left(3^2\right)^3 - \left(2^2\right)^3$ b) $\left(8^{-1}\right)^{-2} - \left(7^{-2}\right)^{-1}$

   c) $\left(5^2\right)^5 - \left(5^2\right)^3$ d) $-\left(4^{-3}\right)^{-1} - \left(5^3\right)^{-1}$

   e) $2^4 + 2^{-2} - 2^0 \times 2^{-2}$ f) $4^2 + 4^{-2} \times 5^{-1}$

   g) $3^{-3} \times \dfrac{1}{3^{-6}} + \left(3^0\right)^3 - 3^{-1}$ h) $\dfrac{\left(3^{-2}\right)^3 \left(4^{-3}\right)^2}{\left(4^2\right)^{-3}\left(3^6\right)^{-2}}$

i) $\dfrac{\left(5^{-2}\right)^3\left(4^4\right)^{-3}}{\left(4^3\right)^{-2}\left(5^3\right)^{-4}}$

j) $\left(\dfrac{\left(7^6\right)^3\left(7^3\right)^{-2}}{7^{14}}\right)^{-1}$

k) $\dfrac{-2^5}{-2^2}$

l) $\left(-1^3\right)^2$

m) $\left(4^2\right)^3$

n) $(-5)^3$

p) $(-4)^4$

q) $-(-4)^4$

10. a) $\left(\dfrac{4}{3}\right)^{-2}$

b) $\left(\dfrac{5}{4}\right)^{-3}$

c) $\left(\dfrac{7}{8}\right)^{-2}$

Note

The square root of an exponent is the exponent divided by **2**.

The square root of x^8 is x^4 since $\left(x^4\right)\left(x^4\right) = x^8$ (the exponents are added)

Practice

11. Find the square root of the following terms.

a) $16y^8$

b) $4y^{32}$

c) $25c^{28}$

d) $64x^{40}$

CHAPTER 12

BEDMAS 2

Refer to Chapter **7**

Example

Evaluate

$$4 \times 3^3 - \{5 + 4(6 - 5) - 8\}$$
$$4 \times 27 - \{5 + 4(1) - 8\}$$
$$108 - \{5 + 4 - 8\}$$
$$108 - 1$$
$$= 107$$

Example

$$-3(-1 + 6)^3 - 3(2 - 5)^2$$
$$= -3(5)^3 - 3(-3)^2$$
$$= -3(125) - 3(9)$$
$$= -375 - 27$$
$$= -402$$

Example

$$-4\{3(-4)^3 + 7 \times 6^2 - 8$$
$$= -4\{3(-64) + 7 \times 36 - 8\}$$
$$= -4\{-192 + 252 - 8\}$$
$$= -4(52)$$
$$= -208$$

Practice

1. $4\{-7 + 4(3^3) - 8(4^2) - 2(-3 + 6)^3 + 8\}$
2. $\dfrac{2\{5(-4^2) + 7(-3 + 7)^2\}}{6\{3 - 2(5^3) - 6(4 - 5)^3\}}$
3. $8(7) + 6(4)^2 - 8 + 4(7)$
4. $6^2(7) - 4(3) + 8(-4 + 6)^2$
5. $5(-4^3) + 7(-6 + 4)^3 + 8$
6. $(8 + 4 - 3)^3 - \{6 + 4(-3 + 7)^2\}$
7. Evaluate if $b = 3$

   a) $b(b + 3)^2$ b) $b(b + 3)$ c) $b\{b(b + 3)^2 + b^2\}$ d) $b(b + 3)(b - 3)$
8. Calculate

   a) $(-6)^2 \div -2$ b) $2^4 \div 2^3$ c) $2(-4)^2 \div 4$ d) $-8 \div (2 - 4)$

CHAPTER 13

POLYNOMIALS 1

In an algebraic expression such as $5x + 3$, $5x$ and 3 are called terms of the expression. $5x$ is a variable term because x is a variable and coefficient of x is 5. The coefficient of a variable is the number in front of the variable (the multiplier of the variable).

- $5x$ means 5 multiplied by x.
- $\frac{1}{2}x$ means $\frac{1}{2}$ multiplied by x

Example 1

Find the coefficient of the following variables:

a) $2x$ b) $\frac{x}{2}$ c) $-\frac{x}{3}$ d) $\frac{3x}{2}$ e) $-4x$ f) $\frac{-3x}{4}$

a) $2x$ The coefficient of x is 2

b) $\frac{x}{2}$ The coefficient of x is $\frac{1}{2}$; $\frac{x}{2}$ is the same as $\frac{1}{2}x$ *Explanation:* $\left(\frac{1}{2}\right)\left(\frac{x}{1}\right) = \frac{x}{2}$

c) $\frac{-x}{3}$ The coefficient of x is $-\frac{1}{3}$

d) $\frac{3x}{2}$ The coefficient of x is $\frac{3}{2}$

f) $\frac{-3x}{4}$ The coefficient of x is $-\frac{3}{4}$

Practice

1. Find the coefficient of the following terms.

a) $\frac{-4xy}{3}$ b) $\frac{2xy}{5}$ c) $-4fg$ d) $\frac{-x}{7}$

Terms

- $4x^2y^3$ is a term (means 4 is multiplied by x^2; multiplied by y^3)
- 2 is the power of x and 3 is the power of y

In the next algebraic expression there are three terms; <u>each term is separated by a plus or minus</u>.
$$4x + 3y - 6z$$

Like terms

- $2x^2y$ and $5x^2y$ are *like terms*, because the power of x is 2 in both terms and the power of y is

1 in both terms. (y means y^1 but the **1** is never shown).

Note:
- $2x^2y$ and $4yx^2$ are *like terms*. The order of the multiplication does not matter. Just as $(5)(4)$ is the same as $(4)(5)$. Both are equal to **20**.
- $4xy$ and $-6xy$ are *like terms*; the minus sign means that the term is negative.
- $2xz$ and $\frac{xz}{3}$ are *like terms* ($\frac{xz}{3}$ means $\frac{1}{3}xz$; $\frac{1}{3}$ is the coefficient of xz)

Unlike terms

- $4x^2y$ and $5x^3y$ are **not** *like terms* because although y is the same power in both terms, x is raised to different powers in the two terms.

The degree of a term is the sum of exponents of its variables

- The degree of $4x^3y$ is **4**, as x is raised to the power of **3** and y is raised to the power of **1** (3+1 = **4**).
- The degree of $10a^6c^5$ is **11**.

Practice

2. Find the like terms in the following.
   a) $3x^2yz, 4xyz, 10x^2yz, 7xy, 4xz$
   b) $5ab, 3abx, 7acb, -6ab, 7ba$
   c) $5x^2z, 3xz, 4ab, 4zx, 5zx$
   d) $-4ab, 3ac, 7ba, 6ca$
   e) $x^2y, 4xy, -7yx, \frac{yx}{2}$

   Find the degree of the following terms
   f) $3xyz$
   g) $4x^4g^4$
   h) $2c^5n^3m$

- An algebraic expression with one term is called a monomial ($3x$).
- With two terms is a binomial ($4xy + 6x$)
- With three terms is a trinomial ($6x + 8d + c$)

Only *like terms* can be added and subtracted

Adding and subtracting polynomials

Examples

1. $4x + 3x = 7x$ (The **4** and **3** are added the same way as integers are added)
                         $4x$ means $x + x + x + x$ just as $4(2) = 2 + 2 + 2 + 2$
2. $4xy - 3xy = xy$ (xy is the same as **1**xy; the one is never written)
3. $-7\,x^2y + 3\,x^2y = -4x^2\,y$
4. $-5.5x + 7x = 1.5x$
5. $\dfrac{8\,x}{7} - \dfrac{3\,x}{5}$

   $= \dfrac{5(8\,x) - 7(3\,x)}{35}$ (refer to the Chapter on LCM)

   $= \dfrac{40\,x - 21\,x}{35}$

   $= \dfrac{19\,x}{35}$

Practice

3. Simplify

a) $-3m + 4m$

b) $-6p + 3p - 8p$

c) $0.3x - 0.6x + 0.9x$

d) $x^3 + 6\,x^2 - 4\,x^3 + 7x^2$

e) $\dfrac{2\,a}{3} + \dfrac{5\,b}{4} - \dfrac{6\,a}{5} + \dfrac{8\,b}{4}$

f) $4xy + 6x^2 - 5xy + 7x^2 + 8$

g) $4b + 7 - (6b - 8)$

h) $2x + 3y - (7x - 4y)$

i) $4xy + 6x^2y - 5xy - (5xy + 7x^2 - 3x^2y)$

j) $\dfrac{2\,m}{3} + \dfrac{3\,n}{5} - \left(\dfrac{3\,m}{5} - \dfrac{7\,n}{3} \right)$

k) $\dfrac{3\,a}{2} + \dfrac{5\,b}{4} - (7a + 8b)$

Multiplying and dividing polynomials

Exponents to the same base in this case x, when multiplied are added algebraically.

Examples

1. $\left(x^3\right)\left(x^2\right) = x^5$
2. $\left(a^{-3}\right)\left(a^6\right) = a^3$
3. $(2\,x^3y^4)\left(4\,x^5y^2\right) = 8x^8y^6$ *(The numbers are multiplied first **4** x **2** = **8**; then the powers are added as in the previous example.)*
4. $(5\,a^3b^2c^2)(-6\,c^{-2}a^{-2}b^{-1}) = -30ab$ ($c^0 = 1$)
5. $(ac)(xy) = acxy$

6. $(axc)(ax) = a^2x^2c$

Multiplying with brackets

- $a(c + d)$ is the factored form
  $= ac + ad$ (**a** is multiplied by each term in the brackets. This is called the distributive property.)

- $ac + ad$ is the expanded form.

- $x(xy + cd + a)$ (factored form)
  $= x^2y + xcd - xa$ (expanded form)

When multiplying two binomials such as $(a + d)(b + c)$, multiply **a** by the two terms of the second binomial and then **d** by the two terms of the second binomial.

$$(a + d)(b + c) = ab + a\,c + db + dc$$

7. $(a+b)(b+d) = ab + ad + b^2 + bd$
8. $(x+y)(x+y)\ \ x^2 + xy + yx + y^2 = x^2 + 2\,xy + y^2$ (xy and yx are like terms)
9. $(a + b)^2 = (a+b)(a+b) = a^2 + ab + ab + b^2 = a^2 + 2ab + b^2$
10. $(a - b)^2 = a^2 - ab - ab + b^2 = a^2 - 2\,ab + b^2$
11. $a(b + c + d) = ab + ac + ad$

Practice

4. Expand (multiply by doing the brackets)
   a) $(x + y)^2$ b) $(x^2y + 2\,y)(x + y)$ c) $4x(xy - cx + cd)$
   d) $(x + 2y)(x + 3y)$ e) $(cxs - d)(cx + d)$

Common factor

In the algebraic expression $a(b + d)$, a is the common factor because $(b + d)$ are both multiplied by a.

Expanding $a(b + d) = ab + ad$.

$ab + ad$ is the expanded form and $a(b + d)$ is the factored form with a as one factor and $b + d$, the other.

Examples

Factor by removing the common factor.

1. $4x+12 = 4(x+3)$
   4 is one factor and $x+3$ the other; when the two are multiplied the result is $4x+12$

2. $\dfrac{30\,x - 5}{5}$

$= \dfrac{5(6\,x - 1)}{5}$

$= 6x - 1$

3. $\dfrac{16\,x - 8}{2}$

$= \dfrac{8(2\,x - 1)}{2}$

$= 4(2x - 1)$

$= 8x - 4$

4. The difference between $-\dfrac{5\,x + 3}{6}$ and $\dfrac{-5\,x + 3}{6}$

The minus in <u>front</u> of $-\dfrac{5\,x + 3}{6}$ expression changes both the signs in the numerator and the expression can be written as $\dfrac{-5\,x - 3}{6}$

Notice the minus is on top of the horizontal bar in the rewritten form, not in front of it . Both versions are equivalent, they are just written differently.

In the second expression, $\dfrac{-5\,x + 3}{6}$, since the minus is on top of the bar, the signs of the numerator stay the same $\dfrac{-5\,x + 3}{6}$;

$\dfrac{-5\,x + 3}{6}$ can be rewritten as $-\dfrac{5\,x - 3}{6}$. Again both are equivalent, just written differently.

To prove that $\dfrac{-5\,x + 3}{6}$ is the same as $-\dfrac{5\,x - 3}{6}$, let us substitute say $x = 2$ in both expressions and the result should be the same

$\dfrac{-5(2) + 3}{6} = \dfrac{-10 + 3}{6} = \dfrac{-7}{6}$ or $-\dfrac{7}{6}$

$-\dfrac{5(2) - 3}{6} = -\dfrac{10 - 3}{6} = -\dfrac{7}{6}$

Both are equal to $-\dfrac{7}{6}$

5. An algebraic expression for one side of a rectangle is $2x + 1$, the other side is 10. Find an expression for the area of the rectangle.

The area of the rectangle is $(2x + 1)10 = 20x + 10$

If the area of a rectangle is $20x + 10$ and one side is $2x + 1$, find the other side.

The other side is $\dfrac{Area}{2\,x + 1} = \dfrac{20\,x + 10}{2\,x + 1} = \dfrac{10(2\,x + 1)}{2\,x + 1} = 10$

Here is a numerical example to explain question **5**.

15 = **5x3 5** = $\dfrac{15}{3}$; **3** = $\dfrac{15}{5}$; **15** is the area (divide the area by one side of a rectangle to

get the other side)

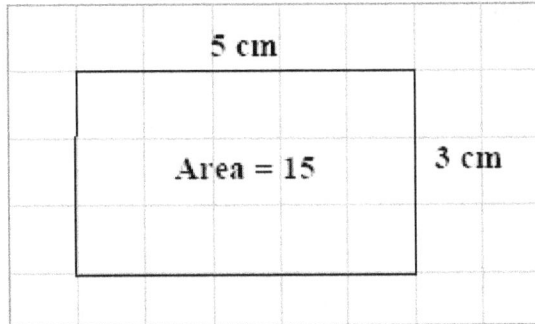

Practice

5. Simplify (remove the common factor first where possible)

a) $\dfrac{16\,x - 4}{2}$

b) $\dfrac{3\,x - 15}{3}$

c) $\dfrac{5\,x + 15}{5}$

d) $\dfrac{3}{8}\left(\dfrac{2\,x}{3} - \dfrac{3\,x}{5} + \dfrac{3}{2} \right)$

e) $\dfrac{3}{2}\left(\dfrac{5\,c}{3} + \dfrac{7}{5} \right)$

f) $3.2(4.2s + 6.1t + 2.3)$

6. Expand and simplify (expand means multiply the brackets and simplify means then collect the *like terms*)

a) $4(8 + 2b) + 4(6 + 3c)$

b) $3(2x - 5y) - 3(3x - 4y)$

c) $3(4x + 3y) - 2(2x - y)$

d) $4(3x + 4y + 6z) - 5(2x + 5y - 3z)$

e) $4s - (2s + 3t) + 7t$

7.
   a) The area of a rectangle is given by the expression $4x + 6$; one side of the rectangle is **2**. Find an expression for the other side.
   b) One side of a rectangle is given by the expression $x + 7$ and the other side is **4**, what is the expression for the area of the triangle.
   c) The area of a triangle is $8x + 12$; one side is $2x + 3$. Find the expression for the other side.

8. $3x + 3y = 9$
   a) Find x when $y = 6$; (express x in terms of y and then substitute for y; refer to the chapter on equations **1**)
   b) Find y when $x = 3$ (express y in terms of x first)

9. $\dfrac{5x}{3} + \dfrac{3y}{2} = 10$; solve for x if $y = \dfrac{2}{3}$

Polynomial equations

In solving equations with fractions, multiply each term by the LCM.

Example

$$\dfrac{x+3}{4} + \dfrac{x-6}{3} = 7$$

The LCM is **12**. Multiply each term by **12**.

$$12\left(\dfrac{x+3}{4}\right) + 12\left(\dfrac{x-6}{3}\right) = 7(12) \qquad \left(\dfrac{12}{4} = 3 \ ; \ \dfrac{12}{3} = 4\right)$$
$$3(x+3) + 4(x-6) = 84$$
$$3x + 9 + 4x - 24 = 84$$
$$7x = 84 - 9 + 24$$
$$7x = 99$$
$$x = 14.14$$

Example

Solve for h
$$2(3h-7) - 7(4h+8) = 31$$
$$6h - 14 - 28h - 56 = 31$$
$$-22h = 56 + 14 - 31$$
$$-22h = 39$$

$$h = -\dfrac{39}{22}$$

$$h = -1\dfrac{17}{22}$$

Practice

10. Solve

   a) $\dfrac{c+4}{3} = \dfrac{c-5}{4}$ b) $\dfrac{y+3}{4} = \dfrac{y+5}{5}$ c) $\dfrac{2x+3}{3} - \dfrac{5x+4}{5} = 5$

11.

   a) $\dfrac{3x+5}{3} - 6x - \dfrac{5}{4} = 6$ b) $3x + \dfrac{7-5x}{4} = \dfrac{15}{4} + \dfrac{5}{2}$

   c) $\dfrac{10}{3} + 6x - \dfrac{9}{4} = \dfrac{15}{6}$ d) $\dfrac{3x-7}{4} = \dfrac{8}{3}$

e) $5 + 6x - 4 = 11$

f) $\dfrac{5x}{4} \quad \dfrac{-6}{5} = 20$

g) $\dfrac{7x}{3} + 4 = \dfrac{12}{7}$

h) $\dfrac{8x}{5} \quad \dfrac{-6}{5} = 3$

i) $7x - 8 - 5 = 3$

j) $6x - 8 = \dfrac{8}{9}$

k) $3x - 8 = \dfrac{14}{3}$

12.

a) $(4xy)(6x^3y^4)$ b) $2abc(4a^3b^3c)$ c) $7xyz(2abc)$ d) $ab\,x^3(2ax)$

Division of polynomials

Examples

1. Simplify $\dfrac{(4)(3)(6)}{(4)(3)}$

    The answer is **6** (since $\dfrac{(4)(3)}{(4)(3)} = 1$)

2. Simplify $\dfrac{4xy}{xy}$

    The answer is **4**

2. Simplify $\dfrac{4a^3b^4}{2ab}$

    The answer is $2a^2b^3$

Note $\dfrac{7 + xy}{xy}$ is not equal to 7; $\dfrac{7+xy}{xy} = \dfrac{7}{xy} + \dfrac{xy}{xy} = \dfrac{7}{xy} + 1$

  7 and *xy* share a common denominator *xy*, as indicated by the line.

Practice

13. Simplify the following.

   a) $\dfrac{6xc}{(6xc)^2}$

   b) $\dfrac{5ab(10a^2b^4)}{5a^4b^6}$

   c) $\dfrac{\left(a^{-3}v^3c^{-4}\right)}{4xy}$

   d) $\dfrac{a^{-3}x^4c^{-2})^{-3}}{\left(a^2x^{-3}c^4\right)^{-2}}$

   e) $(4s^3t^4)^{-3}$

   f) $\dfrac{\left(4x^4v^{-3}a^{-2}\right)^2}{\left(8x^3v^{-4}z\right)^3}$

Chapter Review

1. Simplify by collecting terms

   a) $1\frac{3}{8}a - 2\frac{3}{5}b - 2\frac{3}{4}a + 3\frac{3}{10}b$ b) $-\frac{1}{2}m + \frac{3}{4}n + \frac{1}{3}m - \frac{1}{8}y$

   c) $4.5p + 6.2q - 3.2p + 2.5q$

2.

   a) $3a - 2b - (5a + 6b)$ b) $x + y + z - (5x - 6y + 3z)$ c) $4k + 31 - 6m - (3k + 41 - 9m)$

3. Expand

   a) $3(4a)$ b) $5(4a + 2c)$ c) $(a + 2b)(c + 8d)$

   d) $\frac{3}{4}(4a + 8b + 12c)$ e) $-\frac{2}{3}\left(\frac{3}{4}x + \frac{1}{8}y - \frac{2}{5}z\right)$ f) $\frac{2}{5}\left(\frac{2}{3}a - \frac{3}{4}b + \frac{3}{5}c\right)$

4. Expand and simplify

   a) $m^2(2m + 3m - 8) + m(3m^2 - 6m + 9)$
   b) $5a(3a + 2b - 5) - 3a(2a - 3b + 7)$

5. Factor

   a) $10a + 15$ b) $5ab - 15bc$ c) $18abc - 21ab$
   d) $4x^3 + 6x^2 + 8x$ e) $14x^3 + 21x^2 + 28x$ f) $2x^4y^3 - 4x^3y^2 + 6xy$

6. Solve

   a) $\frac{x+1}{4} = 7$ b) $\frac{x}{4} + \frac{x}{3} = \frac{9}{4}$ c) $\frac{2x}{3} - \frac{3x}{4} = 5$

   d) $\frac{2x+3}{4} + \frac{x-4}{3} = 15$

7. Solve

   a) $\frac{3}{5}(5 + 10x) + 7 = -\frac{2}{3}(9 - 12x)$ b) $6 + \frac{1}{3}(4x - 5) = \frac{3}{7}(3x + 4)$

   c) $\frac{1}{2}(6w - 2) - \frac{3}{5}(12 - 4w) = 7$ d) $2.3(4c - 7) = 5.3(6c + 8)$

   e) $6 - \frac{2}{3}(10 - 5d) = \frac{3}{7}(7d - 14)$

8. A square has an area of $9x^2$. Find an expression for the perimeter.

9. The area of a rectangle is $20x + 10y$. One side is $2x + y$. Find an expression for the perimeter?

10. A rectangular prism has a length of **2x**, a width of **3y** and a height of **3z**. Find an expression that represents the surface area and the volume.

11. The perimeter of a triangle is **20x + 7y**, and one side is **4x + 2y** and the other side is **2x + 3y**, find an expression for the third side.

CHAPTER 14

EQUATIONS 2

Plotting a graph on a grid using a table of values

Plotting the graph of $y = 2x$ using the table of values. Multiply the x value by 2 to find the corresponding y value. x is given a value and the corresponding value for y is calculated as above. x is the independent variable and y is the dependent variable because the value of y depends on the value of x.

X	Y
-2	-4
-1	-2
0	0
1	2
2	4

Graph 1

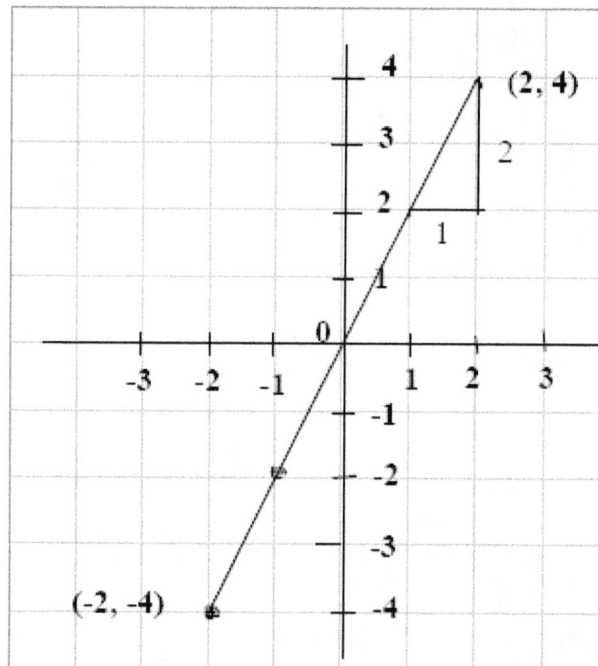

Two points (**-2**, **-4**) and (**2**, **4**) are shown on the Graph 1. The other points fall on the straight line. To draw the graph, any two points can be used. It is not necessary to plot all the points to draw the graph.

The slope of this graph $y = 2x$ is **2**. The slope of a line is equal to $\dfrac{\textbf{rise}}{\textbf{run}}$. The rise of the triangle in the graph is **2** (**2** units vertically) and the run of the triangle is **1** (**1** unit horizontally). Since **2** can be written as $\dfrac{2}{1}$ **2** is the rise and **1** is the run. Hence the slope of the graph is **2**. In this case, the slope is positive because **2** is positive (notice that the graph slopes to the right). If the slope is negative, then the graph slopes to the left as shown in the example below.

Graph 2

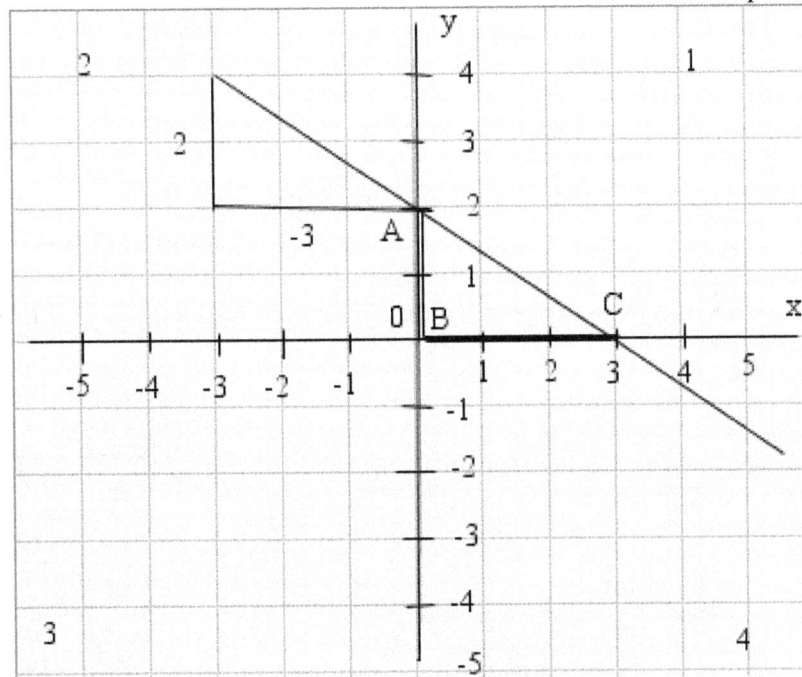

The slope of Graph 2 is $\dfrac{2}{-3} = -\dfrac{2}{3}$

The four quadrants are numbered from **1** to **4** as shown in Graph 2.
- In the first quadrant both x and y are positive.
- In the second quadrant x is negative and y is positive.
- In the third quadrant both x and y are negative.
- In the fourth quadrant x is positive and y is negative.

In Graph 2, triangle **ABC** is in the first quadrant; since x and y are both positive in the first quadrant, the slope should be positive, but the slope is actually negative since it is sloping to the left.

Actually x is negative because starting from the point **C(3, 0)** and move **3** units to the left to **0** (**3** units to the left means -3) and y is positive **2** , so the slope $= \dfrac{\textbf{rise}}{\textbf{run}} = \dfrac{2}{-3} = -\dfrac{2}{3}$

Graph 3

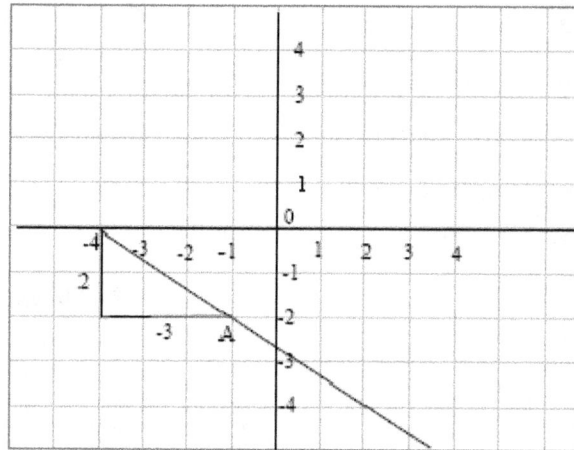

To find the slope of the line in Graph 3 , start at point **A** and move **3** units to the left (-**3**) and then **2** units up (+**2**).

The slope is $\dfrac{2}{-3} = -\dfrac{2}{3}$ Always start at the peak and then move horizontally and then vertically.

Graph 4

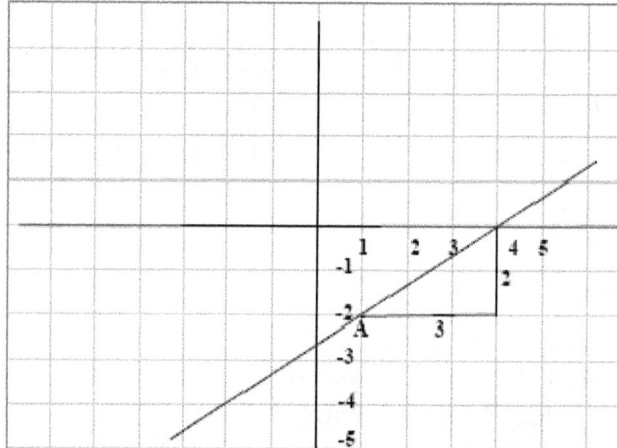

In Graph 4, start at point A and move **3** units to the right (+**3**) and then **2** units up (+**2**)
The slope is $\dfrac{2}{3}$.

Graph 5

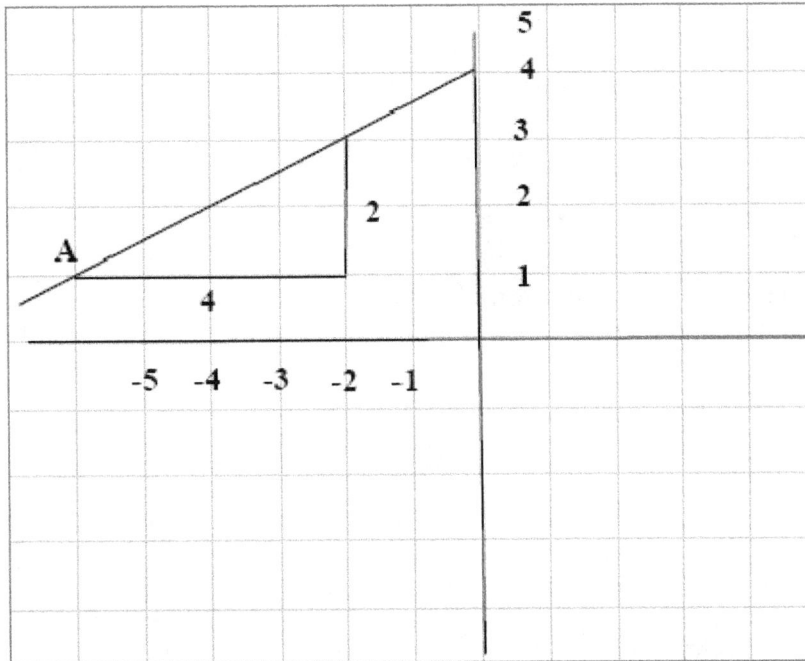

In Graph 5, start at point **A** and move **4** units to the right (+4) and **2** units up (+2)
The slope = $\dfrac{2}{4}$ = $\dfrac{1}{2}$.

The slope of a line can also be calculated as $\dfrac{\textbf{The difference in the y coordinates}}{\textbf{The difference in the x coordinates}}$

Graph 6

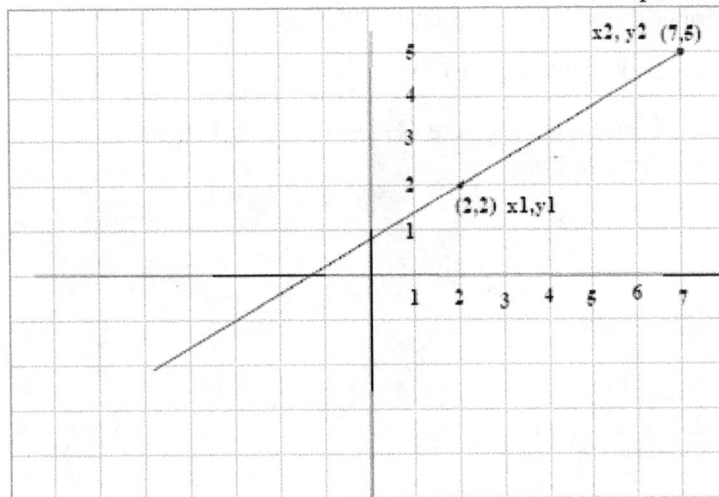

In Graph 6, the slope = $\dfrac{\textbf{y2 } - \textbf{y1}}{\textbf{x2 } - \textbf{x1}}$ = $\dfrac{5 - 2}{7 - 2}$ = $\dfrac{3}{5}$.

Example

Calculate the slope given by the following coordinates.

A(3, 6) B(-4, -2)
$x1\ y1$ $x2\ y2$

The slope $= \dfrac{y2-y1}{x2-x1} = \dfrac{-2 - 6}{-4 - 3} = \dfrac{-8}{-7} = \dfrac{8}{7}$.

Here the **A** coordinates are subtracted from the **B** coordinates, but the same answer can be arrived at by subtracting the B coordinates from the A coordinates as shown below.

The slope $= \dfrac{y1 - y2}{x1 - x2}$

$= \dfrac{6 - (-2)}{3 - (-4)}$

$= \dfrac{6 + 2}{3 + 4} = \dfrac{8}{7}$

<u>The slope of a line is a rate</u>

Since the slope is a comparison between two quantities, it is also the rate. In Graph 7, hours are plotted on the *x* axis and kilometers are plotted on the *y* axis. The comparison is between kilometers versus hours.

- Triangle **ABC** is the rate triangle since $\dfrac{AC}{BC} = \dfrac{rise}{run} = \dfrac{km}{hours}$ which is a rate expressed as kilometers per hour.

- The slope of the line **AB** $= \dfrac{150}{3} = 50$ kilometers per hour which is a rate.

- The steeper the slope, the higher the rate.

- The slope of the line represented by **FG** is $\dfrac{100}{1} = \dfrac{100}{1\ hour} = 100$ kilometers per hour.

- **FG** is steeper than **AB** and so is the rate (**FG** represents **100** kilometers per hour whereas **AB** represents **50** kilometers per hour)

Graph 7

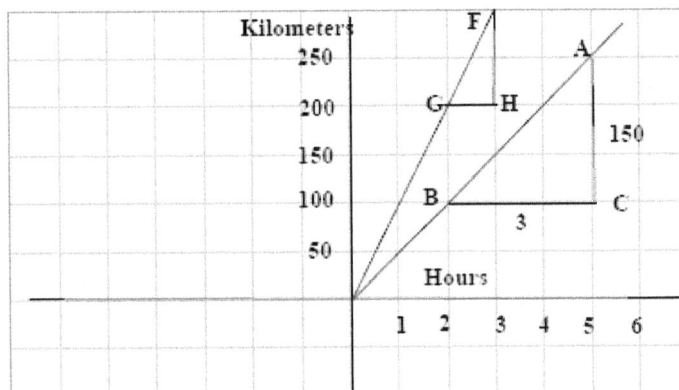

Zero and undefined slopes

Graph 8

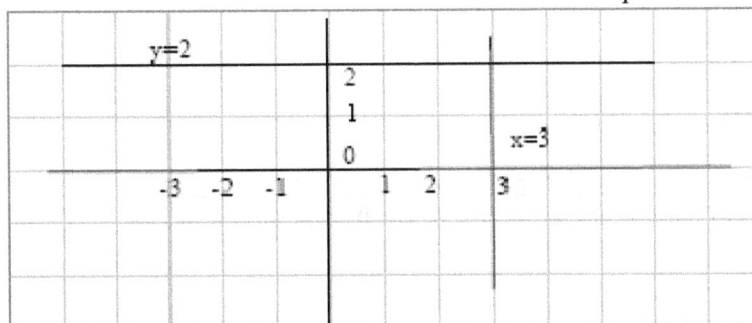

The line represented by the equation $y = 2$ is shown in Graph 8. All points on the line satisfy the equation $y = 2$. The slope of this line is zero because it is horizontal and does not slope to the right or left.

The rise is zero so the slope is $\dfrac{\textbf{zero}}{\textbf{run}}$ = zero.

In Graph 8, the line represented by the equation $x = 3$ has a slope that is undefined because the run = zero.

The slope = $\dfrac{\textbf{rise}}{\textbf{run}} = \dfrac{\textbf{rise}}{\textbf{zero}}$ = undefined.

A vertical line has an undefined slope.

Practice 1

Find the slopes of the following pair of points.

a) **(4, -2)** and **(-6, 4)** b) **(5, -4)** and **(9, 4)**
c) **(-4, 7)** and **(-3, 6)** d) **(-2, 6)** and **(5, -3)**
e) **(4, -3)** and **(-2, 4)** f) **(3, -2)** and **(-5, 4)**

<u>X and Y intercepts</u>

The x intercept is the point at which the graph crosses the x axis and the y intercept is the point where the graph crosses the y axis.

Fig 1

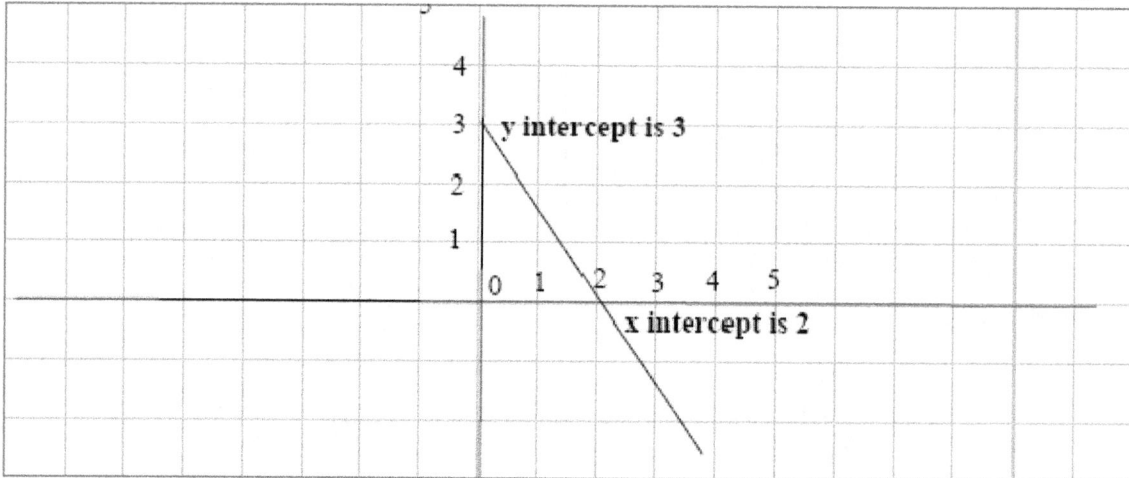

<u>The y intercept occurs when $x = 0$; the x intercept occurs when $y = 0$.</u>
The coordinates of the x intercept in fig **1**, are **(2, 0)** and the coordinates of the y intercept in Fig. 1, are **(0, 3)**

If the x and y intercepts are given the graph can be easily plotted by joining the x intercept and y intercept with a straight line as shown in Fig **1**.

<u>The difference table</u>

Graphs can be plotted using tables by giving values to x and finding the corresponding y values.

Example

Draw the graph of $y = 2x + 1$ using a table of values.

X	1	2	3	4	5
Y	3	5	7	9	11

Difference in y values: 2 2 2 2

The difference in the y values are all **2** (first differences), therefore this graph is linear.
The first differences from the graph are **5 – 3 = 2, 7 – 5 = 2, 9 – 7 = 2, 11 – 9 = 2**

If the first differences are the same, the graph is linear.

Plot the graph using the table of values. The graph will be linear with a *y* intercept of **1**, when *x* = **0**, since *y* is decreasing by **2** as we move to the left .

Practice 2

Create a difference table (first differences) from the table of values in Fig. A, Fig. B, Fig. C and determine whether the graphs are linear or not.

Fig A

X	1	2	3	4
Y	6	9	12	15

Fig B

X	1	2	3	4
Y	3	8	13	18

Fig. C

X	1	3	4	5
Y	4	10	13	16

<u>Standard and slope intercept form of an equation</u>

The standard form of an equation of a line is given by **A*x* + B*y* + C = 0**.
Example

$2x + 3y + 6 = 0$ is an equation of a line in standard form where **A = 2, B = 3** and **C = 6**. **A, B, C** are constants and *x* and *y* are the variables.

An equation can also be written in the slope intercept form.
$y = mx + b$ is the slope intercept form, where *m* is the slope and *b* is the *y* intercept.

Example

$y = 3x + 4$ (**3** is the slope and **4** is the *y* intercept)

To convert an equation in the standard form to the slope intercept form, express *y* in terms of *x*

Example

Rewrite $4x + 3y - 6 = 0$ in the slope intercept form.
$3y = -4x + 6$

$$y = -\frac{4}{3}x + 2; \quad m = -\frac{4}{3} \text{ and } b = 2$$

When the equation is written in this form it is easy to draw the graph without the use of tables. Start with *y* intercept and then draw the slope.

Example

In an equation $y = 3x + 1$, **3** is the slope or *m* and **1** is the y intercept or *b* (**3** is the same as $\frac{3}{1}$)

Fig 2

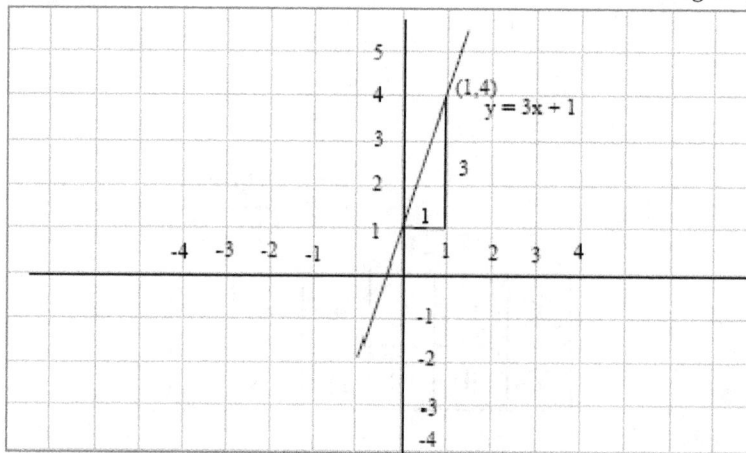

To draw the graph of $3x + 1$, without using the table of values, start with the **y** intercept which is **1**, (refer to fig **2** above). From the **y** intercept, move **1** unit to the right and then **3** units up which brings you to the point (**1, 4**). Join (**1,4**) to the **y** intercept of **1** and extend the line in both directions to get the graph. (Note the slope of **3** means $\frac{3}{1}$; **3** is the rise and **1** is the run)

Practice 3

Draw the following graphs using the slope intercept form.

a) $y = 4x + 4$ b) $y = -3x - 3$ c) $2x - 4y + 5 = 0$ d) $3x - 4y - 8 = 0$

To find the x intercept of an equation in standard form or in the slope intercept form, make $y = 0$

Example

Find the x intercept for the equation $3x + 4y = 8$

Setting $y = 0$, the equation becomes: $3x = 8$; $x = \dfrac{8}{3}$ or $2\dfrac{2}{3}$.

The x intercept is $2\dfrac{2}{3}$

To find the y intercept make $x = 0$

Example

Find the y intercept for the equation $4x + 3y = 10$

By making $x = 0$, the equation becomes: $3y = 10$; $y = \dfrac{10}{3} = 3\dfrac{1}{3}$

The y intercept is $3\dfrac{1}{3}$

Practice 4

Find the x intercept for the following equations.

a) $4x + 3y = 7$ b) $5x + 2y = 8$ c) $y = 4x + 8$

Find the y intercepts for the following equations.

d) $3x + 2y = 8$ e) $x = 2y + 8$ f) $y = 3x + 5$

Parallel and perpendicular lines

Parallel lines have the same slopes, but the slopes of perpendicular lines are negative reciprocals.

Example

Two equations represented by $y = 4x + 6$ and $y = 4x + 10$ are parallel because both lines have a slope of 4.

Two equations represented by $y = \dfrac{3}{4}x + 6$ and $y = -\dfrac{4}{3} + 7$, are perpendicular since the reciprocal of $\frac{3}{4}$ is $\frac{4}{3}$ and the minus in front of $\frac{4}{3}$ makes it a negative reciprocal.

Practice 5

Determine whether the following lines are perpendicular, parallel or neither.

a) $m1 = 3$, $m2 = \frac{1}{3}$

b) $m1 = 2$, $m2 = \frac{-1}{2}$

c) $m1 = \frac{1}{5}$, $m2 = -0.2$

d) $m1 = 0.6$, $m2 = \frac{3}{5}$

e) $m1 = \frac{4}{5}$, $m2 = -0.8$

f) $m1 = c$, $m2 = \frac{-1}{c}$

<u>Writing the equation of a line given a pair of points</u>

To find the equation of a line given a pair of points, first find the slope and then the *y* intercept.

Example

Find the equation of a line passing through **A(4,3)** ; **B(6, -5)**

The slope or $m = \dfrac{y2 - y1}{x2 - x1} = \dfrac{-5-3}{6-4} = \dfrac{-8}{2} = -4$

Using the slope intercept form $y = mx + b$, to find b.
Substitute the values of x, y and m to find b.
The x and y values can be taken either from point **A** or point **B**.
Taking the x and y values from point **A** ; $x = 4$ and $y = 3$

$\qquad y = mx + b$
$\qquad 3 = -4(4) + b$
$\qquad 3 = -16 + b$
$\quad 3 + 16 = b$
$\qquad 19 = b$
$\quad$ or $b = 19$
The equation is $y = -4x + 19$

Note :   In writing the equation of a line the values of m and b are plugged into the equation, but x and y remain as x and y are the variables.

In calculating the slope of a line either use $\dfrac{y2-y1}{x2-x1}$ or $\dfrac{y1-y2}{x1-x2}$ but never $\dfrac{y2-y1}{x1-x2}$

The same order should be used with x and y

Practice 6

Write an equation for each line.

a) Points $(7, 8)$ and $(-4, 5)$
b) Points $(3, 4)$ and $(-4, 5)$
c) Points $(2, -5)$ and $(6, -3)$
d) The slope is undefined and the point $(6, 3)$ is on the line.
e) The slope is zero and the point $(6, 4)$ is on the line.

Examples 1 and 2 are special cases.

Example 1

Write an equation of a line that is parallel to $y = 3$ and passes through the point $(5, 4)$

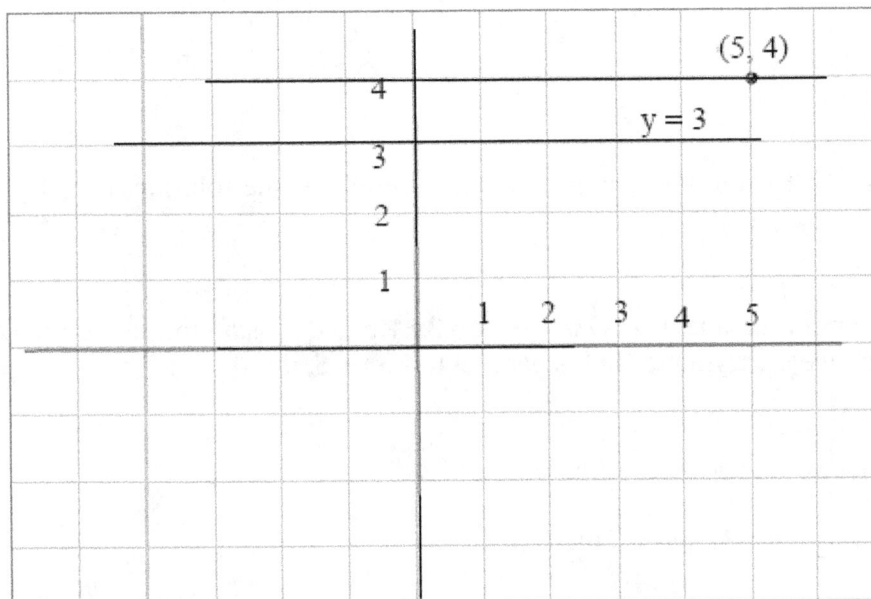

Since the equation we are trying to find out is parallel to $y = 3$, $y = 4$ is the solution. The x value has no effect on the solution.

Example 2

Write an equation of a line that is parallel to $x = 2$ and passes through the point (**3, 2**)

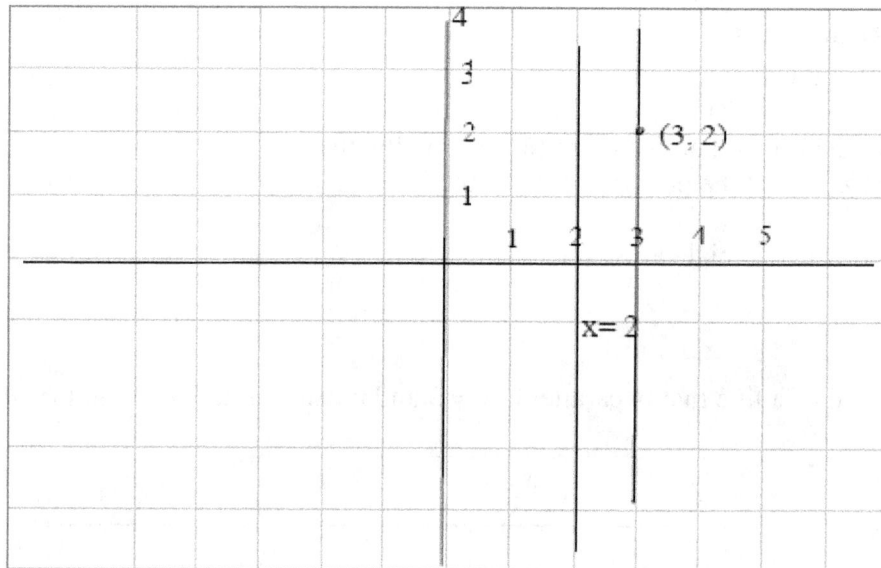

In this case $x = 3$ is the solution. The y value has no effect on the solution.

Example 3

Write an equation for the line that is parallel to - $3x + 5$ and passing through point (**2, 3**). Since the equation that we are trying to find is parallel to - $3x + 5$, the slope of the new equation is also **-3**

The x and y values of the new equation are **2** and **3**

Substituting x, y and m (-3) into $y = mx + b$

$$3 = -3(2) + b$$
$$3 + 6 = b$$
$$b = 9$$

The equation is $y = -3x + 9$

Practice 7

Write an equation of a line that is

a) parallel to - $2x + 5$ and passes through (**3, 4**)
b) perpendicular to - $3x + 7$ and passes through (**5, 6**)
c) perpendicular to $5x + 4$ and has the same y intercept as $4x + 4$
d) parallel to $x = 4$ and passing through the point (**6, 3**)
e) parallel to $y = 3$ and passing through the point (**7, 8**)

The slope of a graph represents rate

Since the slope of a graph is $\dfrac{\textbf{rise}}{\textbf{run}}$ and rate is a comparison of two quantities such as kilometers

per hour or $\dfrac{\textbf{kilometers}}{\textbf{hours}}$, the slope of a graph represents rate. Since the rise corresponds to the *y*

value, kilometers which is the rise is plotted on the *y* axis and hours which is the run is plotted on
the *x* axis.

Example 4

Draw a graph that represents the following: Tiffany earns **8** dollar per hour and Jamie makes **12**
dollars per hour.

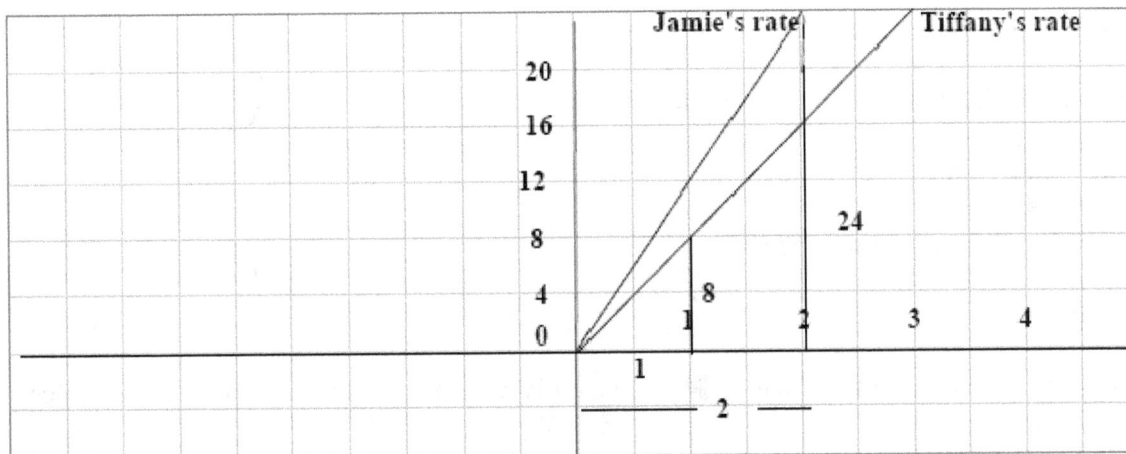

$$\text{Tiffany's rate} \;=\; \frac{\textbf{rise}}{\textbf{run}} \;=\; \frac{8}{1} \;=\; \$8/\text{hour}$$

$$\text{Jamie's rate} \;=\; \frac{\textbf{rise}}{\textbf{run}} \;=\; \frac{24}{2} \;=\; \$12/\text{hour}$$

· This relation is called direct variation because both graphs originate at **0** (there is no *y* intercept)

In this next example, *y* = **3***x* + **5**, there is a *y* intercept of **5**. This relation is referred to as partial
variation.

Here is an example of partial variation.

A taxi charges a fixed amount of $2 plus $3 a kilometer.

The equation to the above problem is **C** = **3***d* + **2** where **C** is the total charge and *d* is the distance
travelled in kilometers. **2** is the *y* intercept which is a fixed amount, which does not change with
the mileage.

What would be the total fare if a person took the taxi for a distance of **15** kilometers?

$C = 3(15) + 2$ (substitute **15** for *d*)
$C = 45 + 2$
$C = 47$
The fare is **47** dollars.

Example 5

Jim's repair shop charges a flat fee of $**30**, plus an hourly rate of $**50** an hour. Write an algebraic equation for this relation.

If the repair took **6** hours, what would be the total charges?

Let the total charges be **C** and the number of hours be *h*.
$C = 50h + 30$ is the equation to the problem.

For a **6** hour repair, substitute **6** for *h* in the same equation
$C = 50(6) + 30$
$C = 300 + 30$
$C = 330$
The total charges equals $**330**.

Example 6

Tidy Car Rentals charges a fixed fee of $**40** and a daily rate of $**20**. Hertz Rent A Car charges $**40** a day.

Write an algebraic equation for the two car rentals.

Graph the two relations.

Which car rental would you choose if you were renting for
a) **4** days   b) **1** day

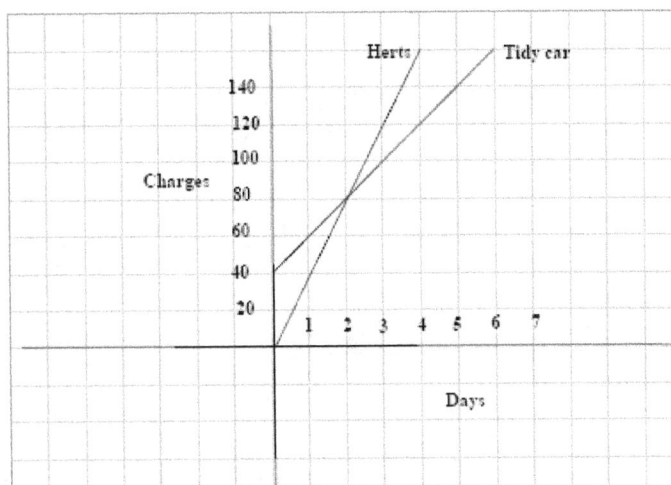

a) The equation for Tidy Car is **C = 20d + 40**
The equation for Hertz is **C = 40d**
Rentals for **4** days for Tidy Car is **C = 20(4) + 40 = 120**
Rentals for **4** days for Hertz is **C = 40(4) = 160**
Tidy Car would be cheaper.

b) From the graph at **2** days, both the rentals would be the same at $**80**.
Under **2** days, Hertz would be cheaper and over **2** days, Tidy Car is cheaper.
Hertz would be cheaper for **1** day.

To calculate at what day the charges for Tidy Car and Hertz would be the same, equate the two equations (**C=20d +40** and **C = 40d**)

$$20d + 40 = 40d$$
$$20d - 40d = -40$$
$$-20d = -40$$
$$d = \frac{-40}{-20} = 2 \text{ days } (\textit{the same as the graph})$$

Practice 8

Find the point of intersection of the following pairs of equations by expressing both equations in the form $y = mx + b$ and then equating them, since both are equal to y.

a) **4x - 3y = 6** and **2x + 4y = 8**
b) **2x + 6y = 8** and **3x + 2y = 6**
c) **x + 2y = 10** and **6x + 4y = 8**

Example 7

A nut wholesaler wishes to mix two grades of nuts. One sells for $8/kg and the other sells for $11/kg. The manager wants to mix the two grades of nuts to get **220** kg selling for **$9.5**/kg. How many kg of each grade must he sell?

Let the number of kilograms of the $8/kg nuts be x, then the number of kilograms of the $11/kg will be **220 - x**

The cost of the nuts = price per kg, multiplied by the number of kilograms
The cost of the $8/kg nuts = **8$x$**
The cost of the $11/kg nuts = **11(220 - x)**
The total cost of the mixture = **9.5(220)**
The cost of the $8/kg nuts plus the cost of the $11/kg nuts must equal the total cost of the mixture.

$$8x + 11(220 - x) \;=\; 9.5(220)$$
$$8x + 2420 - 11x \;=\; 2090$$
$$-3x \;=\; 2090 - 2420$$
$$-3x \;=\; -330$$
$$x \;=\; 110 \text{ kg}$$

There are **110** kg of the $8/kg nuts and **220-110** or **110** kg of the $11/kg nuts.

Cross multiplication

Sometimes cross multiplication is the easiest way to solve equations.

$$\frac{5}{x} = \frac{2}{3}$$

Multiply the numerator of the first fraction by the denominator of the second fraction followed by the equal sign and then multiply the denominator of the first fraction by numerator of the second fraction.

$$15 \;=\; 2x$$
$$x \;=\; \frac{15}{2}$$
$$x \;=\; 7\frac{1}{2}$$

Cross multiplication works well when the unknown is in the denominator.

In the next example the unknown is in the numerator, so no cross multiplication is necessary.

$$\frac{x}{5} = \frac{3}{2}$$
$$x \;=\; \frac{5(3)}{2} \qquad \textit{(move the 5 to the right hand side; no need for cross}$$

$$\textit{multiplying)}$$

Example 8

A chemist has **100** ml of a **45%** acid solution. He wants to reduce the strength to **30%** by adding distilled water. How many litres of distilled water must be added?

The amount of acid in the **45%** solution is **45**ml. This amount does not change.

Let *w* ml represent the amount of distilled water added.

$$\frac{\textbf{ml of acid}}{\textbf{total volume}} = \frac{\textbf{ml of acid}}{\textbf{total volume}}$$

In a **30%** acid solution there are **30** ml of acid to a total solution of **100**ml (the **100**ml consists of **30** ml of acid and **70** ml of water)

$$\frac{45}{100+w} = \frac{30}{100} \quad \textit{cross multiplying (w ml of water are}$$

added to the 100ml of the 45% acid solution)

$$30(100 + w) = 45(100)$$
$$3000 + 30w = 4500$$
$$30w = 1500$$
$$w = 50 \text{ ml}$$

50 ml of water is added to reduce the strength to **30%**

Example 9

How many pounds of coffee beans selling at $3 per pound (lb) should be mixed with **3** lbs of coffee beans selling for $4/lb to get a mixture of coffee beans selling for **$3.80**/lb?

Let the number of lbs of the coffee beans selling for $3/lb be *x*.
To get the dollar value of the beans selling for $3/lb, multiply the number of lbs by the rate = **3x** dollars.

The dollar value of the beans selling for $4/lb = **4(3) = $12**
The dollar value of the two mixtures = **3x + 12**.
The total number of lbs of the two mixtures = ***x* + 3**

equate this ratio to **$3.80** for one lb

$$\frac{\textbf{dollars}}{\textbf{pounds}} = \frac{\textbf{dollars}}{\textbf{pounds}}$$

$$\frac{3x+12}{x+3} = \frac{3.8}{1} \quad \textit{Cross multiplying}$$
$$3x + 12 = 3.8(x + 3)$$
$$3x + 12 = 3.8x + 11.4$$
$$-0.8x = -12 + 11.4$$

$$-0.8x = -0.6$$
$$x = 0.75 \text{ lb}$$

0.75lbs of beans selling for $3/lb.

Alternate method

Equate the dollar value of the two mixtures to the dollar value of the final mixture at $3.80/lb

$$3x + 12 = 3.8(x + 3)$$
$$3x + 12 = 3.8x + 11.4$$
$$-0.8x = -0.6$$
$$x = 0.75 \text{ lb}$$

Practice

1. The cost of a large pizza is $8 plus **0.50** cents per topping.

   a) Write an equation for the cost of the pizza in the form of $y = mx + b$
   b) Find the cost of a pizza with **6** toppings.
   c) The cost with no toppings.

2. Parcel post charges $4 plus $1 per kg. Courier express charges $3 plus $1.5 per kg.

   a) Write an equation for both services.
   b) Graph both equations on the same set of axes.
   c) What does the point of intersections mean?
   d) To ship a parcel weighing **5** kg, which service would be cheaper?

3. A car rental agency has two plans.
   Plan A : $50 a day plus **10c** a km.
   Plan B : $40 a day plus **20c** a km.

   a) When will the two plans cost the same?
   b) When is plan A better than plan B?
   c) For driving **80** km a day, which plan should be used?

4. John the plumber charges a flat fee of $60, plus an hourly rate of $60. Plumber Jack charges only an hourly rate of $90.

   a) When will the two plumbers charge the same?
   b) If a job was going to take **6** hours, which plumber would be cheaper?

5. Bill wants to start a lemonade business. The cost to start the business is $2. It costs him **6** cents a cup to make the lemonade. He sells the lemonade for **9** cents a cup.

   a) Write an expression for his costs.

b) Write an expression for his revenue.

c) How many cups of lemonade must he sell to break even?

6. A chemist has **100** ml of a **40%** acid solution. She wants to reduce the strength to a **30%** acid solution by adding distilled water. How many ml of distilled water must be added to reduce the strength to **30%** acid?

7. How many pounds of coffee selling at $**2.50**/lb must be mixed with **4** lb of coffee selling at $**4.00**/lb to get a mixture selling at $**3.50**/lb

8. How many litres of a **30%** acid solution must be mixed with **20** L of a **17%** acid solution to get a **25%** acid solution?

Interest

The bank pays interest on money in accounts held by individuals.

If a person has $**1000** in a bank account and the interest rate is **3%** a year, then the interest on

$**1000** is **1000** x $\dfrac{3}{100}$ = $**30**

The interest rate quoted, is the rate per year.

Multiply the interest rate by the amount to get the interest.

9. Mary wants to invest $**10,000** partly in an investment that pays **5%** per year and the rest of the $**10,000** in an investment that pays **7%** per annum. If she wants her total investment to earn an interest of $**600** in one year, how much should she invest in each investment?

10. A rectangle has an area of **30**cm^2. If the width is **3** cm and the length is increased by **3** cms, what is the new perimeter and area of the rectangle?

CHAPTER 15

GEOMETRY 2

<u>Parallel lines and a transversal and more on angles</u>

PQ‖RS and **AB** is the transversal (a line intersecting two or more other lines)

Fig 1

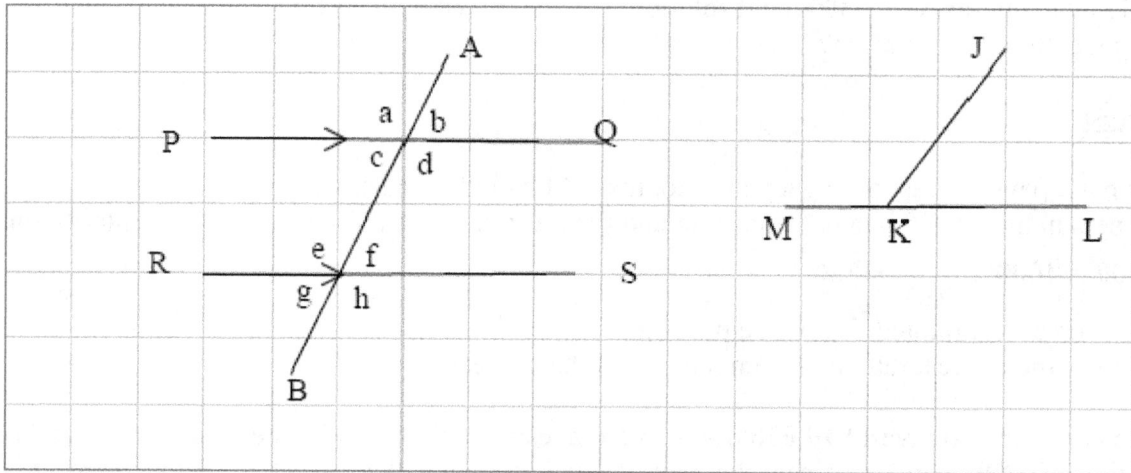

Refer to Fig. **1**

$\angle b = \angle f$; $\angle d = \angle h$; $\angle a = \angle e$; $\angle c = \angle g$ (corresponding angles)

$\angle c = \angle f$; $\angle d = \angle e$ (alternate angles)

$\angle a = \angle d$; $\angle b = \angle c$; $\angle e = \angle h$; $\angle f = \angle g$ (vertically opposite angles)

$\angle d + \angle f = \mathbf{180}$ degrees ; $\angle c + \angle e = \mathbf{180}$ degrees   (interior angles on the same side of the transversal)

In Fig. **1** to the right $\angle JKL + \angle JKM = \mathbf{180°}$ (**180°** stands for **180** degrees.)

Fig 2

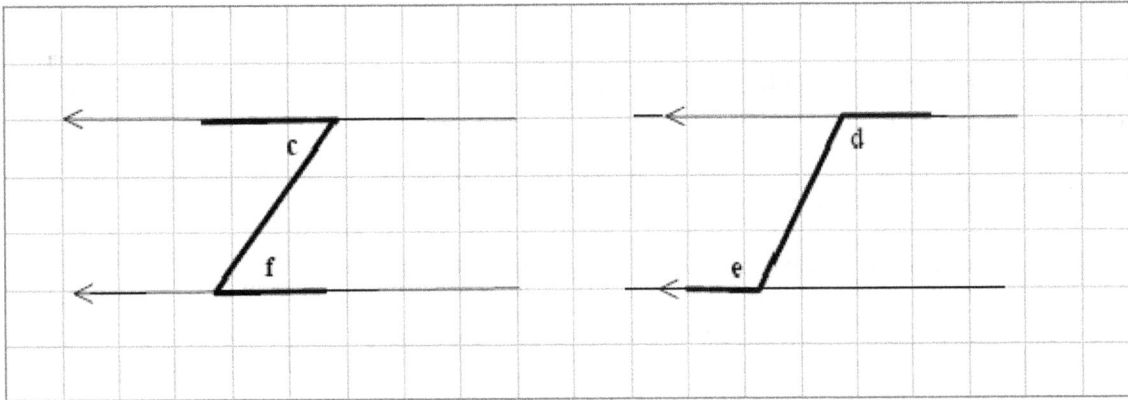

In Fig. 2 the alternate angles *c* and *f*; *d* and *e* are shown in the form of the letter **Z**.

Fig 3

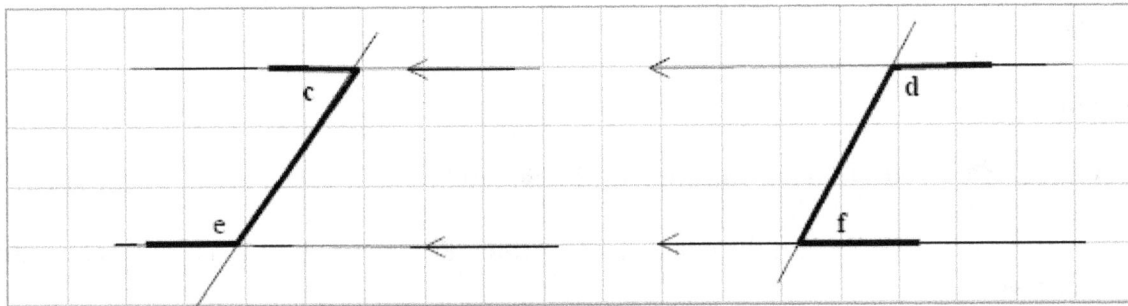

In Fig. 3, the interior angles on the same side of the transversal *c* and *e*; *d* and *f* are shown in the form of the letter **C**.

Fig 4

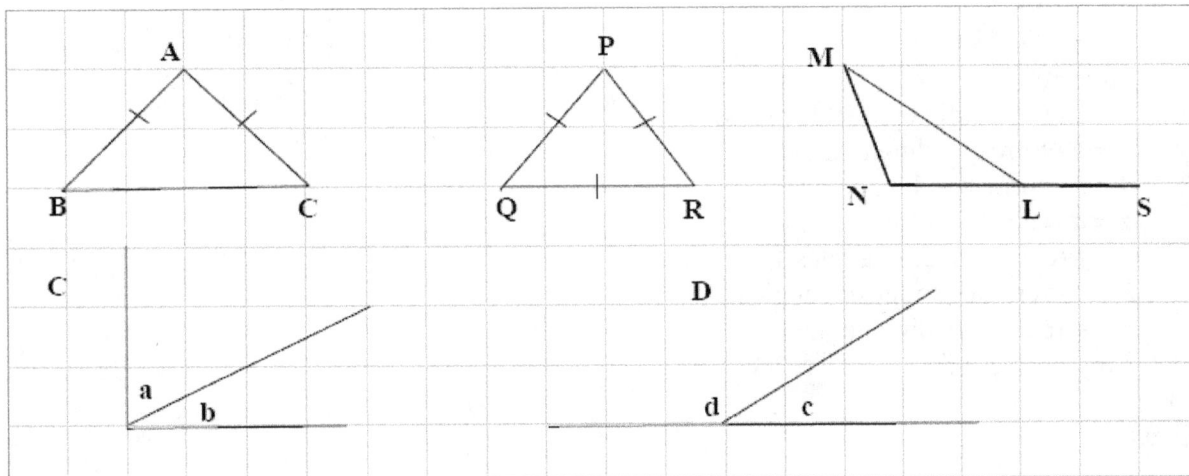

In Fig. 4, △**ABC** is isosceles **AB = AC** ; ∠B = ∠C (The base angles in an isosceles triangle are equal).
△**PQR** is equilateral ∠P = ∠Q = ∠R = **60°**.

In △ **MNL**, the exterior angle **MLS** is equal to the sum of the interior opposite angles **M** and **N**.
If **MLS = 130°** then **M + N = 130°**

<u>The sum of the angles in a triangle is equal to **180** degrees</u>

In Fig. 4 **C**, *a* and *b* are complementary angles because their sum is **90°**.

In Fig. 4 **D**, *c* and *d* are supplementary angles as their sum is **180°**

Example

In Fig.4, △ **ABC** is isosceles. ∠B and ∠C are the base angles. If ∠C = **30°**, find the other angles.

Then ∠B must be equal to **30°**. ∠A + ∠B +∠C = **180°** ; but ∠B + ∠C = **60°**
∠A = **180 - 60 = 120°**

Example

In △ **ABC**, if ∠A = **100°**, find the other angles if AB = AC

The sum of the other angles must be equal to **180 - 100 = 80°**. Each angle must be equal to **40°**
∠B = ∠C = **40°**

Example

Refer to Fig.1. If *a* = **130°**, find *b*, *c*, *d*, *e*, *f*, *g* and *h*

    a + *b* = **180°**
      b = 180 -130
      b = **50°**
     a = *d* (vertically opposite angles)
    a = *e* (corresponding angles)
    d = *h* (corresponding angles)
    a = *d* = *e* = *h* = **130°**
    b = *f* (corresponding angles)
    b = c (vertically opposite angles)
    c = *g* (corresponding angles)
    b = *f* = c= *g* = **50°**

Example

In Fig. 4, if ∠MLS = **120** , and ∠M = **50** , find ∠N

$$\angle MLS = \angle M + \angle N$$
$$120 = 50 + \angle N$$
$$\angle N = 120 - 50$$
$$\angle N = 70°$$

Example

Fig 5

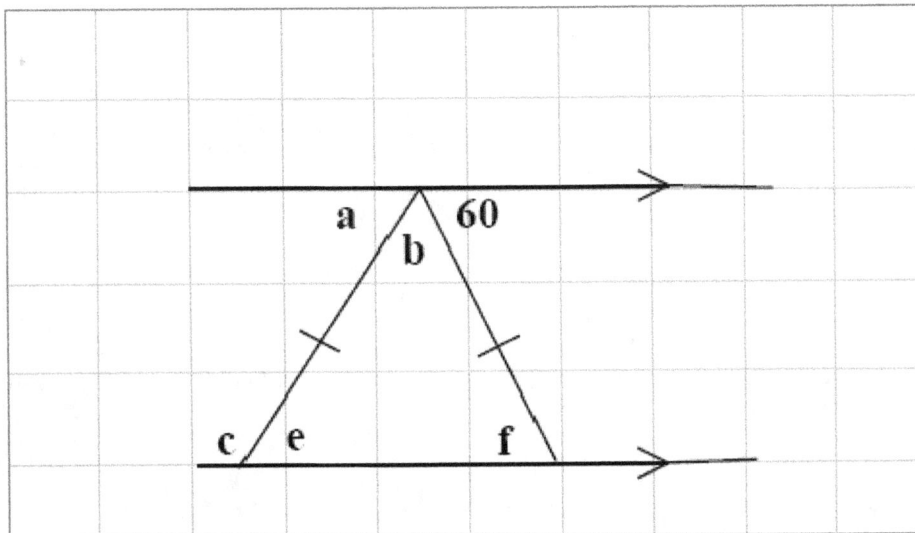

Find the measure of a, b c, e and f, in Fig. 5

f = 60○ (alternate angles Z formation)
f = e = 60○ (base angles of an isosceles triangle)
b = 180 -($e + f$)
b = 180 - 120
b = 60○
c = $b + f$ (Exterior angle + sum of the interior opposite angles)
c = 60 + 60
c = 120○
$a + c$ = 180○ (interior angles on the same side of the transversal = **180○**; **C** formation)
$a + 120 = 180$;
   a = 180 - 120
   a = 60○
   a = 60°, b = 60°, c = 120° , e = 60°, f = 60°

Example

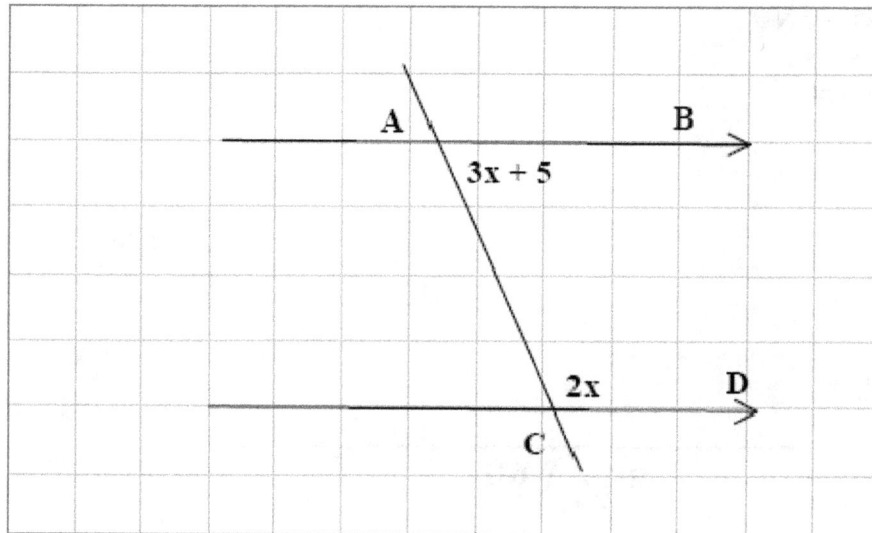

Find ∠**BAC** and ∠**DCA**

3x + **5** and **2**x are the interior angles on the same side of the transversal (C form). Their sum = **180°**.

$$3x + 5 + 2x = 180°$$
$$5x = 180 - 5$$
$$5x = 175$$
$$x = 35°$$
$$∠BAC = 3(35) + 5 = 110°$$
$$∠DCA = 2(35) = 70°$$

Practice

1. a) Find the values of *a*, *b* and *c*.

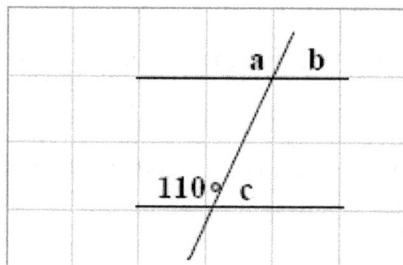

b) Identify the angles corresponding to the Z and C formations in Fig. 1 on page 146

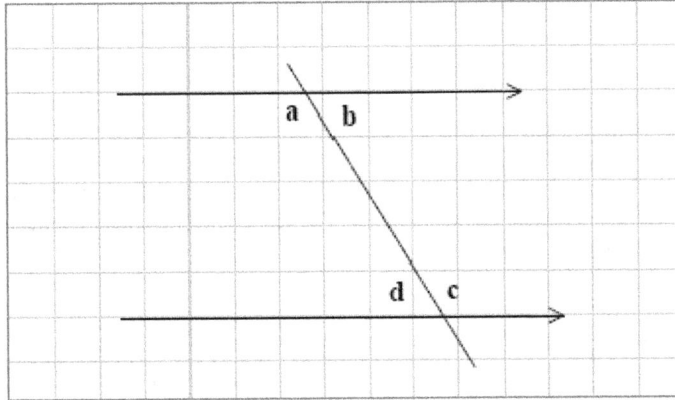

Fig 1

2. Find the measure of each letter in the triangles in Fig. 2 (refer to Geometry 1)

Fig 2

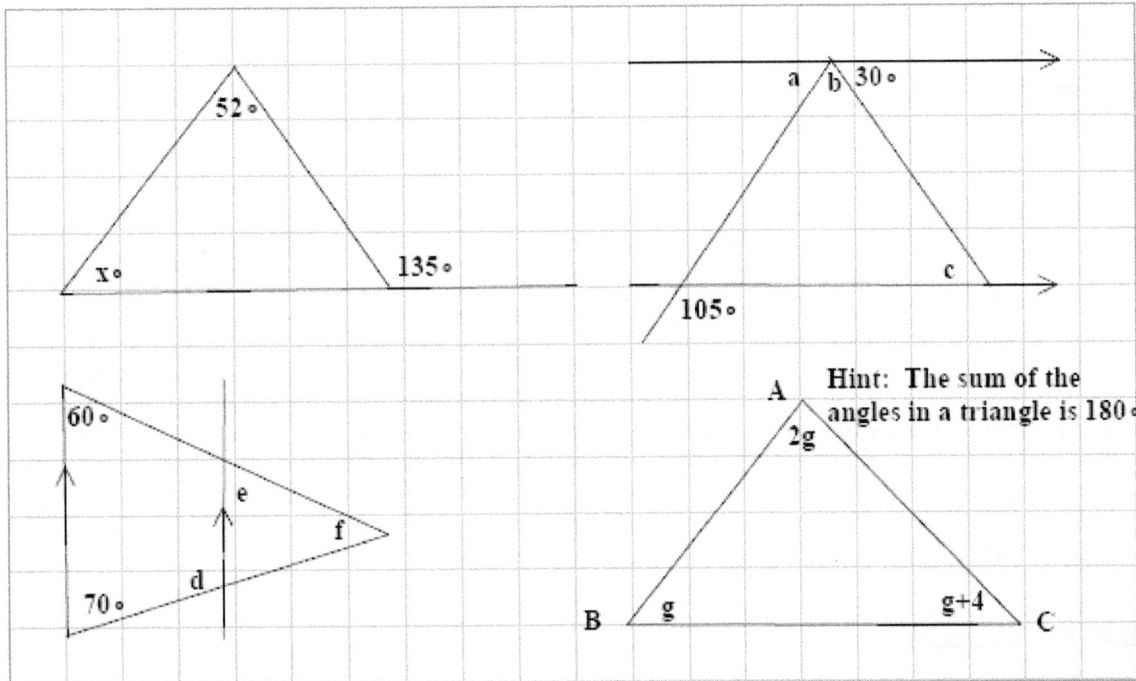

Hint: The sum of the angles in a triangle is 180°

3. Find the measure of each lettered angle in Fig. 3.

Fig 3

Perimeter is 28 cm

A

$3x + 2$

$2x$

B

$2x + 1$

C

Find the length of each side

$3x$ / $2x$

a / b

d / c

f / e

Find the measure of each angle

y / g

40°

$2g$

z

Find the measure of g , y and z

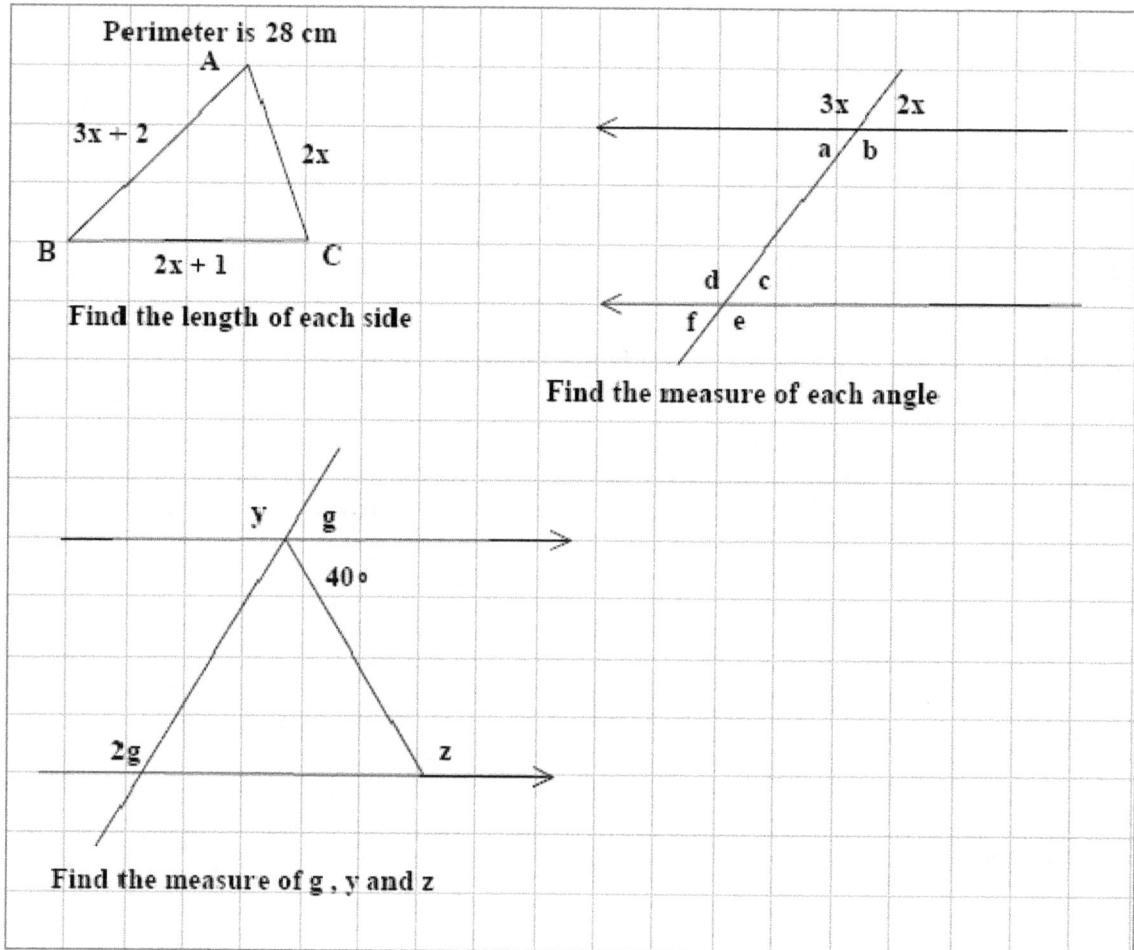

Polygons are figures with three or more sides.

Properties of polygons

The median of a triangle is the line joining a vertex to the midpoint of the opposite side.
The three medians meet at the centroid (**C**) in the figure on the next page. The centroid divides each median in the ratio of **1:2**, so that **AC:CM = 2:1**

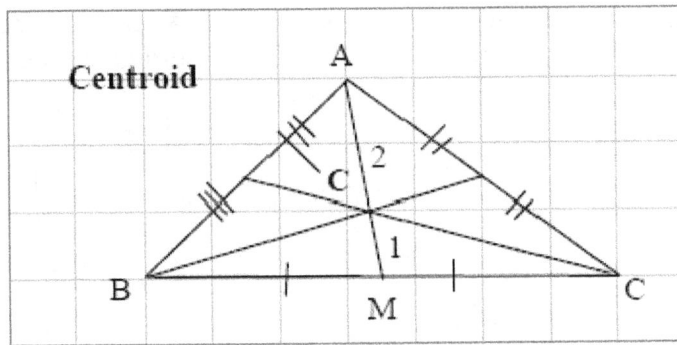

The altitudes of a triangle meet at the orthocentre (**O**). The altitude is obtained by dropping a perpendicular from the vertex to the opposite side.

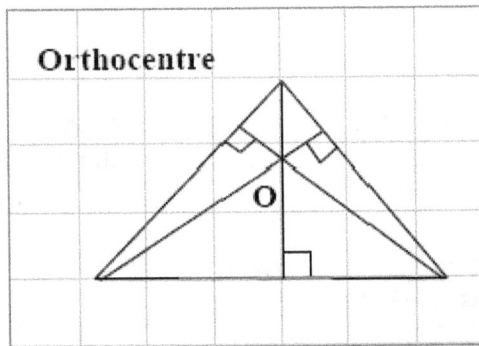

The perpendicular bisectors of the sides of a triangle meet at the circumcentre. The circumcentre is the centre of a circle that passes through all the vertices of a triangle. The circle is called the circumcircle.

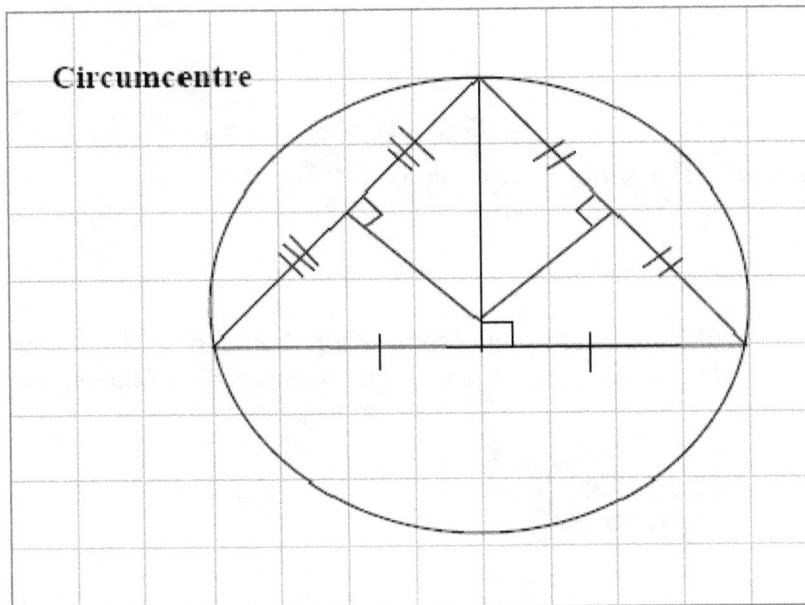

The bisectors of the interior angles of a triangle meet at the incentre. A circle can be drawn with centre I and touches the sides at one point. The circle is called the inscribed circle.

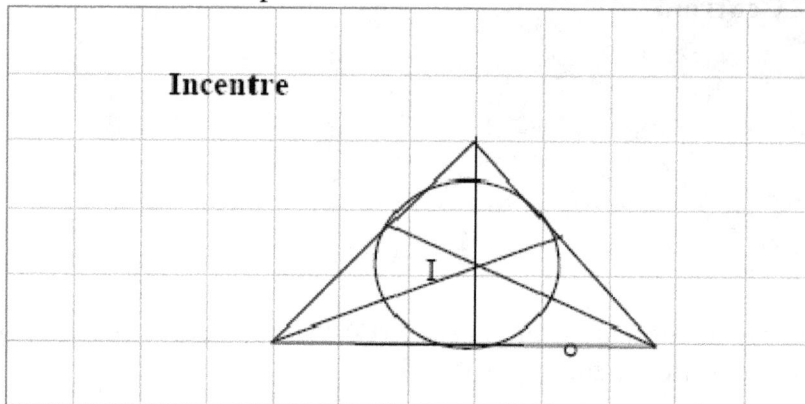

Incentre

Angles of Polygons

A polygon with all sides equal is called a regular polygon. The interior angles (angles at the vertices) of a regular polygon are equal. To measure the sum of the interior angles of a regular hexagon **ABCDEF**, (see Fig. 2 on page 150)

Calculate the sum of the angles of the **6** triangles **AOB**, **BOC** etc.
The total number of degrees of **6** triangles = **6(180)**
The angle at the centre **0 = 360∘**
The sum of the angles at the vertices = **6(180) – 360** or **180n - 360** where n stands for the number of sides of the polygon.
Taking out the common factor of **180**, the expression is equal to **180(n - 2)**.

<u>The sum of the interior angles of a regular polygon = **180(n - 2)**</u>

Example

The sum of the interior angles of an octagon = **180(8 - 2) = 180 x 6 = 1080°**
If the octagon is regular then each interior vertex angle = $\dfrac{1080}{8}$ = **135°**

Areas of polygons

The apothem is the perpendicular drawn from the centre of a polygon to each side. (See Fig. 2).
If the apothem and one side of a regular polygon are given, then the area of the regular polygon can be calculated.

<u>The apothem is the height of the triangle</u>

Example

Find the area of a regular octagon with an apothem of **10** cm and side length of **8.28** cm.

The area of one triangle is $\dfrac{10(8.28)}{2} = 41.4$ cm^2

The area of the regular octagon is **8** x **41.4** = **331.2** cm^2

The example above will help you understand the formula which follows for calculating the area of a regular polygon.

The formula for calculating the area of a regular polygon is $\dfrac{\textbf{perimeter x apothem}}{2}$

The area of a polygon = $\dfrac{P(a)}{2}$

Applying this formula to the worked out example, the area of the octagon = $\dfrac{8.28(8)(10)}{2}$

= **331.2** cm^2 (**8.28** x **8** is the perimeter of the octagon)

Fig. 2

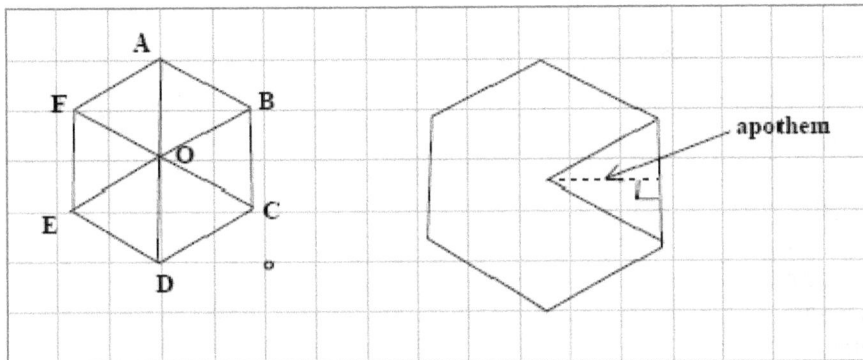

Volumes of three dimensional objects (prisms)

The volume of a regular hexagonal prism is equal to the area of the base x height. The area of the base is the area of the hexagon.

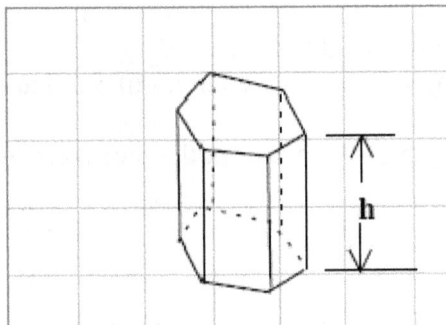

Example

Calculate the volume of a regular hexagonal prism with a height of **8** cm an apothem of **4** cm and a side length of **3.5** cm.

The area of the base is equal to the area of the regular hexagon which is equal to
$$\frac{Pa}{2} = \frac{3.5(6)(4)}{2} = 42\text{cm}^2 \quad \text{(The perimeter is equal to 3.5 x 6)}.$$
The volume of the regular hexagonal prism is equal to the area of the base x height
$$= 42 \times 8 = 336\text{cm}^3$$

A cylinder can be considered to be a circular prism and the volume of the prism is equal to the area of the base x height.
The area of the base is a circle. The area of a circle is Πr^2. (Refer to Geometry 1.) The volume of a cylinder is $\Pi r^2 h$.

Example

Calculate the volume of a cylinder with a radius **4** cm and a height of **6** cm.

The volume of the cylinder is $\Pi(4^2)(6) = 301.59$ cm³

Practice

4. Find the vertex angle to the following.
   a) A regular hexagon.
   b) A regular pentagon.
   c) A regular decagon.
   d) A regular octagon.

5. Find the area of the following regular polygons.
   a) A regular octagon with a side length of **8** cm and an apothem of **6** cm.
   b) A regular pentagon with a side length of **6** cm and an apothem of **5** cm.
   c) A regular decagon with a side length of **5** cm and an apothem of **4** cm.

6. Calculate the volume of the following 3D figures.
   a) A cylinder with a radius of **5** cm and a height of **8** cm.
   b) A regular octagon with an apothem of **4** cm, a side length of **5** cm and a height of **8** cm.
   c) A regular pentagon with a side length of **9** cm, an apothem of **8** cm and a height of **11** cm.

Pyramids

In prisms all the opposite faces are parallel, but with pyramids, the faces meet at a point.

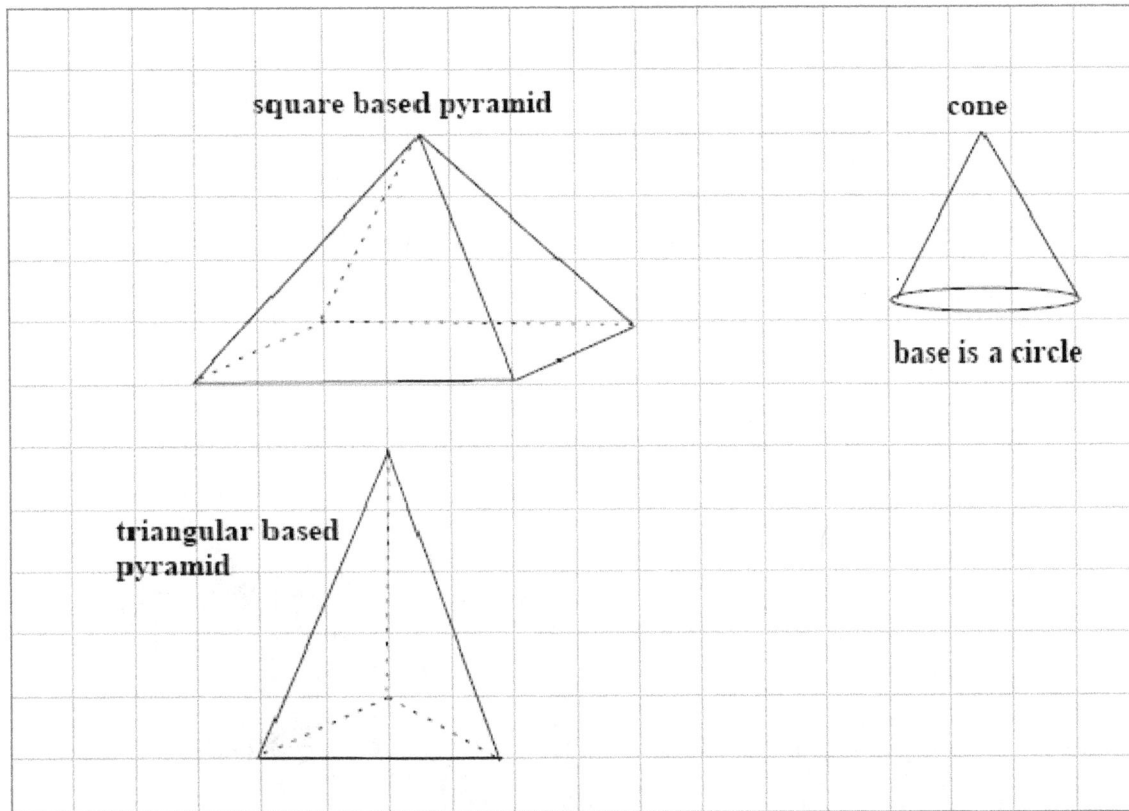

square based pyramid

cone

base is a circle

triangular based pyramid

The volume of a pyramid is $\frac{1}{3}Ah$ ($\frac{1}{3}$ x the area of the base x height).

The volume of a cone is $\frac{1}{3}\Pi r^2 h$ (area of the base is Πr^2).

Example

Find the volume of a regular pentagonal pyramid with an apothem of **5** cm, a side of **6** cm and a height of **8** cm.

The area of the base is the area of a pentagon with side **6** cm and an apothem of **5** cm.

The area of the base is $\frac{Pa}{2}$

The volume of the pyramid is $\left(\frac{1}{3}\right)$ the area of the base x height.

$= \frac{1}{3}\left\{\frac{pah}{2}\right\}$

$= \left(\frac{1}{3}\right)\frac{6(5)(5)(8)}{2} = 200$ cm³   (P = 6(5) , a = 5)

Example

Calculate the volume of the triangular pyramid in Fig. 2, on page 153

Fig 2

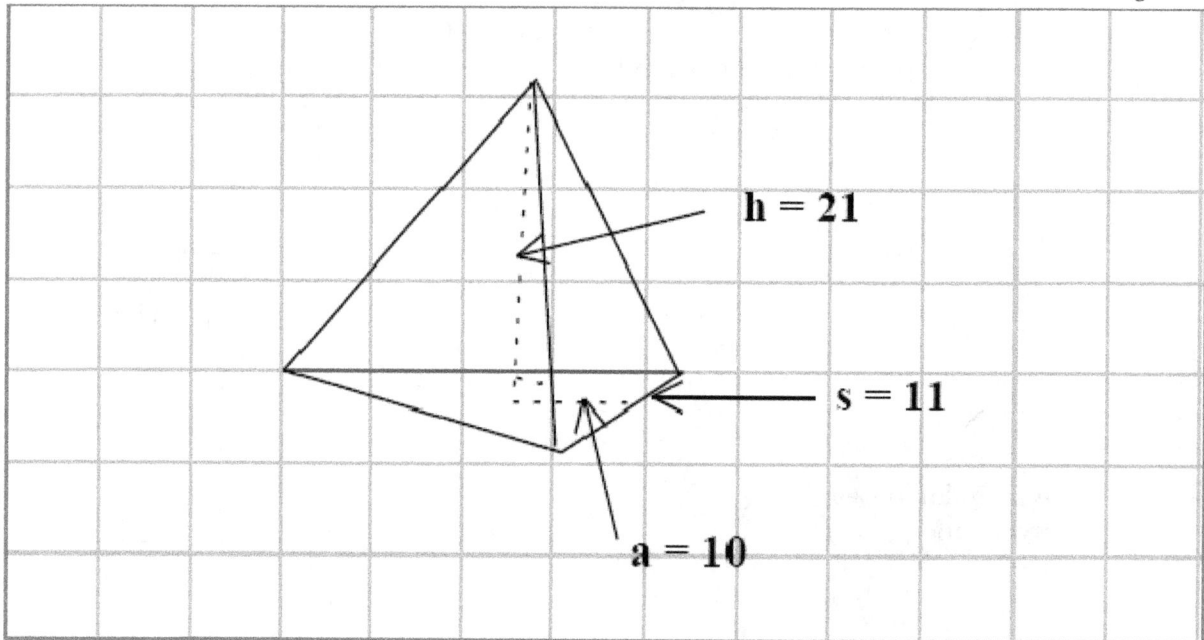

Volume = $\left(\dfrac{1}{3}\right)\dfrac{3(11)(10)}{2}$ x21 = **1155** cubic units.   {Perimeter = **3(11)** }

Spheres

The volume of a sphere is $\dfrac{4}{3}\Pi r^3$

If the radius of a sphere is **3** then the volume = $\dfrac{4}{3}\Pi(3)^3 = \dfrac{4}{3}\Pi(27) = 113.1$

Example

Ice cream is sold to stores in cylindrical containers. If a container is **20** cm high and has a radius of **9** cm, how many scoops of ice cream can be sold if a scoop of ice cream is a sphere of diameter **4** cm.

The volume of the cylindrical container is $\Pi r^2 h = \Pi(81)(20) = $ **5089.38** cm^3.

The volume of the scoop is $\dfrac{4}{3}\Pi(2)^3 = $ **33.51** cm^3.

The number of scoops = $\dfrac{\textbf{Volume of the cylinder}}{\textbf{volume of a scoop}} = \dfrac{\textbf{5089.38}}{\textbf{33.51}} = $ **151** scoops.

Practice

7. Four golf balls with a diameter of **5** cm are to be packaged into a cylinder end on end.
   a) What is the minimum volume of this cylinder?
   b) How much free space does the cylinder have?

8. A rectangular box with a length of **12** cm, a width of **5** cm and a height of **6** cm is designed to package business cards. The dimensions of the memo cards are length = **6** cm and width = **5** cm and a thickness of **1** mm. How many business cards can fit in the box?

9. Calculate the volume of a cone with a diameter of **6** cm and a height of **7** cm.

10. A hexagonal pyramid has a volume of **108** cm^3. If the side length is **4.5** cm and the height is **6** cm. Find the apothem.

11. The volume of a cone is **100.54** cm^3. Calculate the height of the cone if the radius is **4** cm.

12. A triangular based pyramid has a volume of **26.25** cm^3. If the height is **7** cm and the apothem is **2.5** cm, find the length of each side.

Proportion

Direct and inverse proportion

The relation between speed, distance and time is given by the formula, **speed = $\dfrac{\text{distance}}{\text{time}}$**

If time is constant, then speed is directly proportional to distance.
If a car travels **20** kilometers in one hour then the speed is **20** km/h.
If the car travels **40** kilometers in one hour then the speed is **40** km/h.
Therefore speed is directly proportional to distance.

Speed is inversely proportional to the time

If the distance is constant and the time is doubled, the speed is halved.
If the speed is **20** km/hour and the time is doubled to **2** hours the speed becomes **10** km/hour $\left(\dfrac{20}{2}\right)$
Therefore, speed is inversely proportional to time.

In the formula V = $\Pi r^2 h$, the volume is directly proportional to the square of the radius and directly proportional to the height.
If the radius is doubled, then the volume is quadrapuled; (2^2 = 4; increased by **4** times).
If the radius is tripled (**3** times), the volume is multiplied by **9**; (3^2; square of the radius).
Since the height is directly proportional to the volume, if the height is doubled, then the volume is also doubled.
If the radius and height are both doubled, then the volume is multiplied by $(2^2)(2) = 8$.

The effect of changes in linear measurements on the volumes of solids

Example

The effect of changing the radius of a sphere and the corresponding increase in volume is determined by the formula for the volume of a sphere.

The volume of a sphere is $V = \frac{4}{3}\Pi r^3$. $\frac{4}{3}\Pi$ is constant, but the volume varies as the radius cubed (r^3).

If the radius is doubled, the volume will be increased by a multiple of r^3 or $2^3 = 8$ times, since the volume is directly proportional to the radius cubed.
If the radius is tripled then the volume will be increased a multiple of 3^3 or **27** times.
If V is proportional to $\boldsymbol{r^3}$
($\boldsymbol{r^3}$ is proportional to **V** ; $\sqrt[3]{\boldsymbol{r^3}}$ is proportional to $\sqrt[3]{\boldsymbol{V}}$; r is proportional to $\sqrt[3]{\boldsymbol{V}}$)

If the volume is increased by a multiple of **8**, the radius will increase by a multiple of $\sqrt[3]{\boldsymbol{8}}$ or **2**
If the volume is increased by a multiple of **27**, then the radius will increase by a multiple of $\sqrt[3]{\boldsymbol{27}} = 3$

These check out with the two examples above.

Example

The volume of a cylinder is **500 cm^3**. If the radius is quadrupuled (increased four times), what will the new volume be?

The volume will be **500 x 4^2 = 500 x 16 = 8000** cm^3

Example

The volume of a cylinder, $\mathbf{V = \Pi r^2 h}$
The volume is directly proportional to the height. If the height is doubled then the volume is doubled provided r is the same as before.
The volume is proportional to the square of the radius. If the radius is tripled then the volume is increased by a multiple of $\mathbf{3^2}$ or **9** provided h stays the same as before.
If *h* and *r* are both increased then the volume increases by a multiple of the product or **h** and $\mathbf{r^2}$

Example

If the height of a cylinder is doubled and the radius tripled, by what multiple will the volume be increased?

The volume is increased by a multiple of **2 x 3^2 = 18**. The volume is increased by **18** times.

Example

The volume of a cylinder is **750** cm³. If the radius is doubled and the height is tripled, what is the new volume of the cylinder?

The new volume = **750** x 2^2 x **3** = **750** x **12** = **9000** cm³   (replace r with **2** and *h* with **3**)

Example

The volume of a regular hexagonal pyramid is given by V = $\dfrac{1}{3}\left(\dfrac{Pa}{2}\right)h$

$\left(\dfrac{Pa}{2}\right)$ is the area of the hexagon or the area of the base.
P is the perimeter, which a side length multiplied **6** in this case.

The side length and the apothem are directly proportional to each other. If the side length is doubled then the apothem is also doubled. If the apothem is tripled then the side length is also tripled.

Since the calculation for the area of the base includes the side length x apothem, if the side is doubled, the area of the base is multiplied by **4**, or if the apothem is tripled then the area of the base increases by a multiple of **3** x **3** = **9**

Example

Two regular pyramids have the same height and same base configuration. One has a volume of **300 cm³** and an apothem of **3** cm. The other pyramid has an apothem of **9** cm. What is the volume of the other pyramid?

Since the second pyramid has an apothem of **9** cm or three times the apothem of the first pyramid, the volume of the second pyramid is **3** x **3** times the volume of the first. (Since the apothem is increased by three times then the side is also increased by three times, so the volume is increased by **3** x **3**)

The volume of the second pyramid is **9** x **300** = **2700** cm³

Since the height is the same for both pyramids, the height does not affect the result of the calculation.

Practice

13. The radius of a cylinder is **3** cm. If the radius is increased to **9** cm and the height is unchanged, what percentage does the volume increase by?

14. The volume of a sphere is **250** cc. Its volume is increased to **750** cc. What is the percentage increase of the radius?

15. An octagonal pyramid's height is tripled.
   a) What is the percentage increase of the volume?
   b) If the side and height are doubled, what is the percentage increase in volume?

16. A cube's side is doubled, what is the percentage increase in volume?

17. The length and width of a rectangular prism are halved. What is the percentage decrease in volume?

Surface areas of three dimensional objects

Surface area of a cylinder

The surface area of a cylinder is $2\Pi r^2 + 2\Pi rh$ ($2\Pi r^2$ is the area of the top and bottom and $2\Pi rh$ is the curved surface area)

To simplify the calculations $2\Pi r^2 + 2\Pi rh$ can be simplified to $2\Pi r(r + h)$

The surface area of a rectangular prism is $2(LW + WH + HL)$

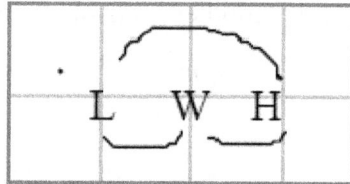

An easy way to remember this combination is to write **L W H** and then take two at a time as shown by the line in the diagram.

The surface area of a cone

The surface area of a cone is $\Pi r^2 + \Pi rl$ or $\Pi r(r + l)$ where l is the slant height.

The slant height $l = \sqrt{r^2 + h^2}$ (pythagorean theorem)

If the radius and height are given then l can be calculated.

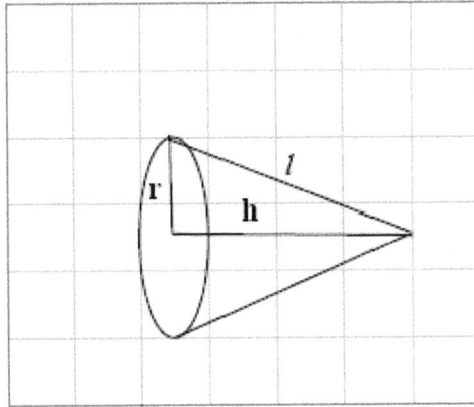

Example

Find the surface area of a cone with radius **3** cm and height **5** cm.

The slant height l is $\sqrt{3^2 + 5^2} = \sqrt{9 + 25} = \sqrt{34} = 5.83$

$l = 5.83$

The surface area of the cone $= \mathbf{\Pi r^2 + \Pi rl} = \Pi(3^2) + \Pi(3)(5.83)$
$= \Pi(9) + \Pi(17.49) = 28.27 + 54.94 = \textbf{83.21 cm}^2$

Example

Find the surface area and volume of the square based pyramid in Fig. 1 on page 159.
$s = \mathbf{6}$ cm ; $a = \mathbf{5}$ cm ; $h = \mathbf{7}$ cm

There are four triangles congruent to **ABC** and a square **BEDC**
To find the area of triangle **ABC** we have to know the length of **AF** which is the height of triangle **ABC**.

$AF^2 = h^2 + a^2$
$AF^2 = 7^2 + 5^2 = 49 + 25 = 74$
$AF = \sqrt{74} = 8.6$
$AF = 8.6$

Area of triangle is equal to $\dfrac{AF(BC)}{2} = \dfrac{8.6(6)}{2} = \textbf{25.8 cm}^2$ *(BC = s = 6)*

The area of the four triangles $= 4(25.8) = \textbf{103.2 cm}^2$.
The area of the square $= 6 \times 6 = \textbf{36 cm}^2$.
The total surface area of the square pyramid is $103.2 + 36 = \textbf{139.2 cm}^2$.

The volume of the pyramid $= \dfrac{1}{3}$**the area of the base x height.**

The volume of the pyramid $= \dfrac{1}{3} 36 \; x \; 7 = \textbf{84 cm}^3$

Fig **1**

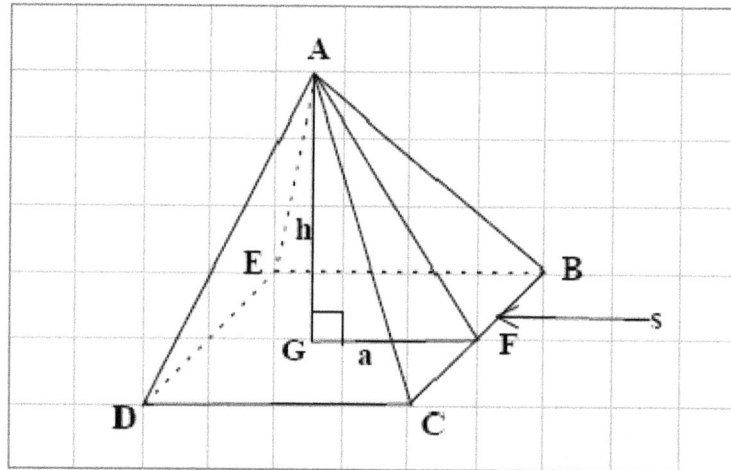

Chapter Review

1. The surface area of a sphere is equal to $4\Pi r^2$. If the surface area of a sphere is **50.26**, what is the volume of the sphere?

2. A square based pyramid has a height of **7.4** cm, a side of **6** cm and an apothem of **3** cm, find the surface area and the volume.

3. What is the interior angle of a regular octagon?

4. A regular octagon has a side length of **6** cm and an apothem of **7.24** cm. Calculate the area.

5. A regular hexagonal pyramid has an apothem of **4.33** cm and a side length of **5** cm. If the height of the pyramid is **8** cm, calculate the volume of the pyramid.

6. A cone has a radius of **3** cm and a height of **6** cm. A square based pyramid with a side of **6** cm and the same height as the cone. Which has the larger volume and by how much?

7. A square based pyramid has a side length of **8** cm. What is the apothem for this pyramid?

8. Samantha ordered soil for her gardens. The soil was poured in her backyard in the form of a cone **2** m in diameter and **0.5** m high. She has to pour this soil into her **8** gardens which measure **50** cm by **40** cm. What will be the height of the soil in her gardens?

9.

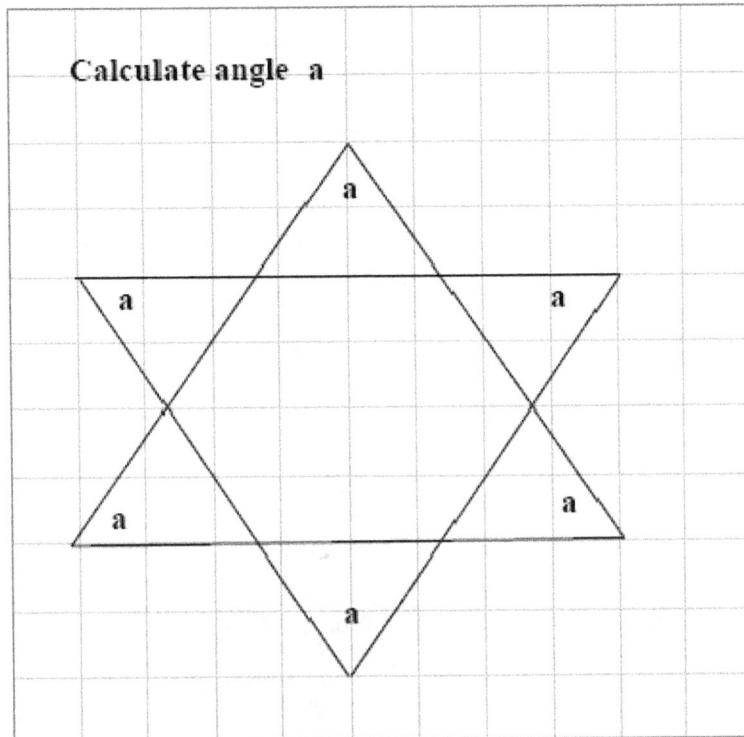

Calculate angle a

The geometry of packaging

Minimum surface area of a cylinder for a given volume.

For a cylinder with a given volume, there is a relationship between radius and height that yields the minimum surface area.
This occurs when the diameter equals the height of the cylinder.

$$V = \Pi r^2 h \qquad (h = 2r)$$
$$V = \Pi r^2 (2r)$$
$$V = 2\Pi r^3$$
$$r^3 = \frac{V}{2\Pi}$$
$$r = \sqrt[3]{\frac{V}{2\Pi}}$$

Maximum volume of a cylinder for a given surface area.

For a cylinder with a given surface area, there is a relationship between radius and height that yields a maximum volume.

This occurs when the diameter equals the height of the cylinder.

$$SA = 2\Pi r^2 + 2\Pi rh$$

$SA = 2\Pi r(r + h)$ ($2r = h$, when max volume occurs)

$SA = 2\Pi r(r + 2r)$

$SA = 2\Pi r(3r)$

$SA = 6\Pi r^2$

Example

Find the dimensions of a cylindrical can that will use the least amount of metal if the volume of the can is **1000** ml.

The minimum surface area occurs when $r = \sqrt[3]{\dfrac{V}{2\Pi}}$

$r = \sqrt[3]{\dfrac{1000}{2\Pi}}$

$r = \sqrt[3]{159.15}$

$r = 5.42$ cm

The radius of the can is **5.42** cm and the height is **10.84** cm.   ($h = 2r$)

Example

Sam has **550cm^2** of metal to build a cylindrical juice can to hold the maximum amount of juice. What are the dimensions of the juice can?

The maximum volume for a given surface area occurs when

$SA = 6\Pi r^2$

$r^2 = \dfrac{550}{6\Pi}$

$r = \sqrt{\dfrac{550}{6\Pi}}$

$r = 5.40$cm $h = 2r = 10.8$cm

The radius of the can is **5.40** cm and the height is **10.80** cm.

Practice

1. A cylindrical juice can must hold **600** ml of juice.
   a) What are the dimensions of the can so that the least amount of metal is used?
   b) Determine the surface area of this can.
   c) If aluminum costs **0.012** cents per **cm^2**, what is the cost of this can?

2. Solve for *r*

   a) $A = \Pi r^2$ b) $V = \dfrac{4}{3}\Pi r^3 (6)$ c) $SA = 6\Pi r^2$

3. A glass manufacturer wants to use the minimum amount of glass for making a drinking glass (cylindrical with no top) holding **200** ml. What is the dimension of the glass?

4. A drinking glass has a surface area of **75** cm² (the drinking glass has no top). What are the dimensions of the glass so that a maximum volume can be obtained? (Derive the formula and then use it in the question.

<u>Minimum surface area of a rectangular prism for a given volume</u>

For a rectangular prism with a given volume, there is always a length, width and height in which a minimum surface area occurs.
The minimum surface area occurs when the length equals the width equals the height.
$V = l$ x w x h Since $l = w = h$
$V = l^3$
$l = \sqrt[3]{V}$

<u>Maximum volume for a given surface area of a rectangular prism</u>

For a rectangular prism with a given surface area there is always a length width and height in which a maximum volume occurs.

The maximum volume occurs when the prism is a cube. (length = width = height)

The area of each side is l^2. The total surface area is $6l^2$
$SA = 6l^2$
$l = \sqrt{\dfrac{SA}{6}}$
Example

Sid has **1000** cm² of metal to build a rectangular box that must accommodate the largest possible volume. Determine the dimensions of the box.

The greatest volume for a given surface area occurs when
$l = \sqrt{\dfrac{SA}{6}}$
$l = \sqrt{\dfrac{1000}{6}}$
$l = \textbf{12.91}$ cm

The box must be **12.91** cm in length, width and height.

Example

A rectangular glass fish tank is built to have a capacity of **1** litre.
  a) Find the dimensions of the tank that will use the least amount of glass.
  b) Calculate the area of glass used.

For a given volume, the least surface area occurs when $l = \sqrt[3]{v}$

  a) $l = \sqrt[3]{v}$
     $l = \sqrt[3]{1000}$
     $l = 10$ cm
     $l = w = h = 10$ cm

  b) Each face of the cube is **10** x **10** = **100** cm^2

The area of glass used is **5** x **100** = **500** cm^2 (The top is open)

Practice

5. A rectangular swimming pool is to be built of metal and holds a capacity of **5 m^3**. If the least amount of metal is to be used, calculate the surface area of the swimming pool.

6. A rectangular box with a surface area of **16067 cm^2** is to be designed for maximum volume What are the dimensions of the box?

7. If **1250 cm^2** was shaped into a cylindrical prism or a rectangular prism for a maximum volume. Which shape would produce the greater volume and by how much?

8. A laundry company wishes to box **8** litres of detergent into a rectangular box using the least amount of cardboard for the maximum volume. What are the dimensions of the box?

CHAPTER 16

EQUATIONS 3

Simultaneous equations

So far we have dealt with single equations, but many problems can be solved by using simultaneous equations. When two sets of conditions are given by two equations then simultaneous equations should be used. Simultaneous equations can be solved graphically as in the next example

Example

Solve graphically:

$$-x + y = 4 \quad \textbf{(1)}$$
$$-2x + y = 2 \quad \textbf{(2)}$$

Rewriting both equations in the form $y = mx + b$ so that they can be graphed.

$$y = x + 4 \quad \textbf{(3)}$$
$$y = 2x + 2 \quad \textbf{(4)}$$

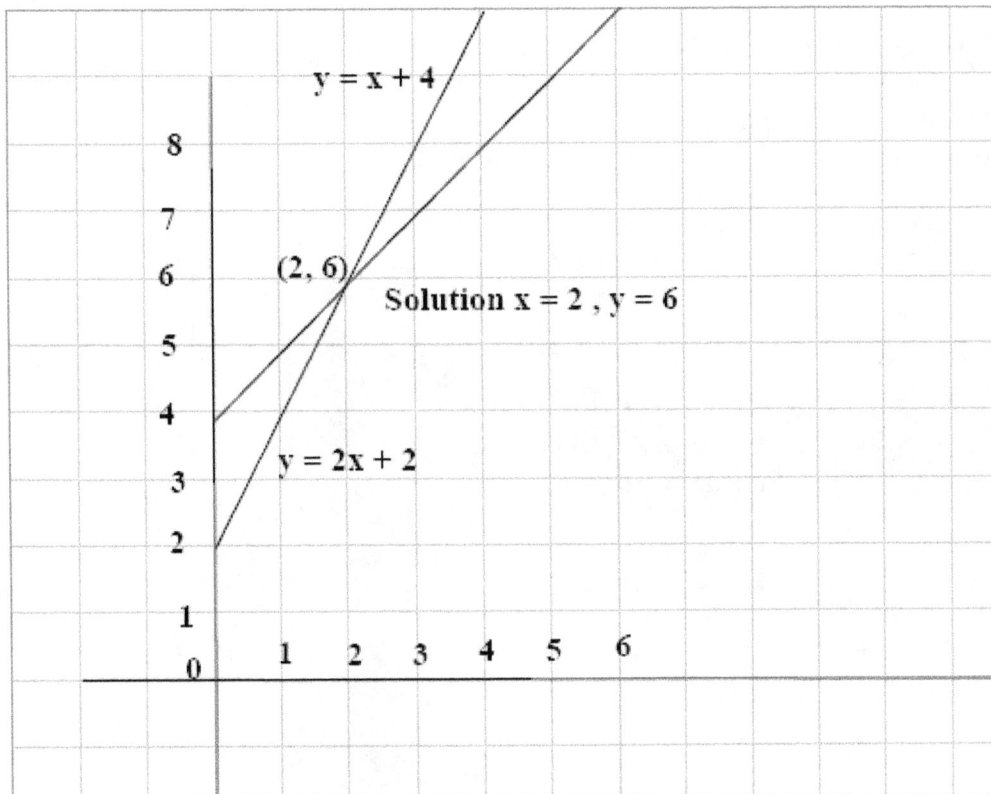

The point of intersection is the solution, since that point is common to both equations.

Simultaneous equations can also be solved algebraically by substitution as in Example 1 and 2.

Example 1

Substitution

$$x + y = 10 \quad (1)$$
$$3x + 4y = 20 \quad (2)$$

To solve this pair of equations by substitution, we isolate either x or y in one equation and substitute this value in the other equation.

From equation 1

$$
\begin{aligned}
x + y &= 10 \\
y &= 10 - x \quad \textit{Substituting this value of y in equation 2.} \\
3x + 4(10 - x) &= 20 \\
3x + 40 - 4x &= 20 \\
-x &= 20 - 40 \\
-x &= -20 \quad \textit{(Changing signs)} \\
x &= 20
\end{aligned}
$$

(Changing signs is the same as multiplying both sides of the equations by **-1**)

Substitute this value of x in either equation 1 or 2
Substituting **20**, for x in equation 1

$$
\begin{aligned}
20 + y &= 10 \\
y &= -20 + 10 \\
y &= -10
\end{aligned}
$$

Example 2

$$2x + y = 8 \quad (1)$$
$$5x + 3y = 10 \quad (2)$$

From equation 1

$$y = -2x + 8$$

Substituting for *y* in equation 2

$$5x + 3(-2x + 8) = 10$$
$$5x - 6x + 24 = 10$$
$$-x = 10 - 24$$
$$-x = -14$$
$$x = 14$$

Substituting for *x* in equation 1

$$2(14) + y = 8$$
$$28 + y = 8$$
$$y = 8 - 28$$
$$y = -20$$

Practice 1

Solve by substitution

a) $x + 3y = 2$ b) $x - 3y = 5$ c) $2x + 3y = 6$ d) $3x + 2y = 8$
   $3x + 2y = 4$ $x + 2y = 6$ $5x + 6y = 9$ $2x + 4y = 10$

Simultaneous equations can be solved by elimination as in examples 3 and 4

Example 3

Elimination

This method is most commonly used.

$$3x + 2y = 10 \qquad (1)$$
$$4x + 5y = 8 \qquad (2)$$

The idea is to make *x* or *y* the same in both equations and then eliminate the equal variable. Multiply Equation 1 by 4 (all the terms) and multiply equation 2 by 3

$$3x + 2y = 10 \qquad (1) \text{ x } 4$$
$$4x + 5y = 8 \qquad (2) \text{ x } 3$$
$$12x + 8y = 40 \qquad (3)$$
$$12x + 15y = 24 \qquad (4)$$

(Subtract Equation 4 from Equation 3; change the signs of all the terms in Equation 4). Rewriting Equation 3 and the new Equation 4

$$12x + 8y = 40 \qquad (3)$$
$$-12x - 15y = -24 \qquad (4) \text{ } (adding \text{ } equations \text{ } 3 \text{ } and \text{ } 4;$$
$$\text{-----------------} \qquad \qquad +12x \text{ } and - 12x = 0)$$
$$-7y = 16$$
$$y = \frac{16}{-7} = -\frac{16}{7}$$

Page 166

Substituting for y in Equation 1

$$3x + 2\left[-\frac{16}{7}\right] = 10$$

$$3x - \frac{32}{7} = 10$$

$$3x = \frac{32}{7} + 10$$

$$3x = \frac{32+70}{7}$$

$$3x = \frac{102}{7}$$

$$x = \frac{102}{3(7)}$$

$$x = \frac{102}{21} = \frac{34}{7} = 4\frac{6}{7}$$

$$x = 4\frac{6}{7}; \quad y = -\frac{16}{7} = -2\frac{2}{7}$$

Example 4

$$\frac{x}{3} + \frac{y}{4} = 1 \qquad (1)$$

$$\frac{x}{2} + \frac{y}{3} = 4 \qquad (2)$$

To eliminate the denominators, multiply Equation 1 by 12 which is the LCM.

$$12\left[\frac{x}{3}\right] + 12\left[\frac{y}{4}\right] = 12$$

$$4x + 3y = 12 \qquad (3)$$

Multiply each term in Equation 2 by 6

$$6\left[\frac{x}{2}\right] + 6\left[\frac{y}{3}\right] = 6$$

$$3x + 2y = 24 \qquad (4)$$

Rewriting Equation 4 below Equation 3

$$4x + 3y = 12 \qquad (3) \quad \text{Multiply by } \mathbf{2}$$
$$3x + 2y = 24 \qquad (4) \quad \text{Multiply by } \mathbf{3}$$

To eliminate y, multiply Equation 3 by **2** and Equation 4 by **3**

$$8x + 6y = 24 \qquad (5)$$
$$9x + 6y = 72 \qquad (6)$$

Subtract Equation 6 from Equation 5 (change signs of all the terms in Equation 6) and then add the two equations as in the Example 3.

Re writing Equation 5

$$8x + 6y = 24 \qquad (5)$$
$$-9x - 6y = -72 \qquad (6)$$

$$-x = -48$$
$$x = 48$$

Substituting for **x** in Equation 5

$$8(48) + 6y = 24$$
$$384 + 6y = 24$$
$$6y = 24 - 384$$
$$6y = 360$$
$$y = \frac{360}{6} = 60$$
$$x = 48 \; ; \; y = 60$$

<u>Note</u>

When equations have different slopes, there is one solution as in the above examples and shown graphically in the figure below.

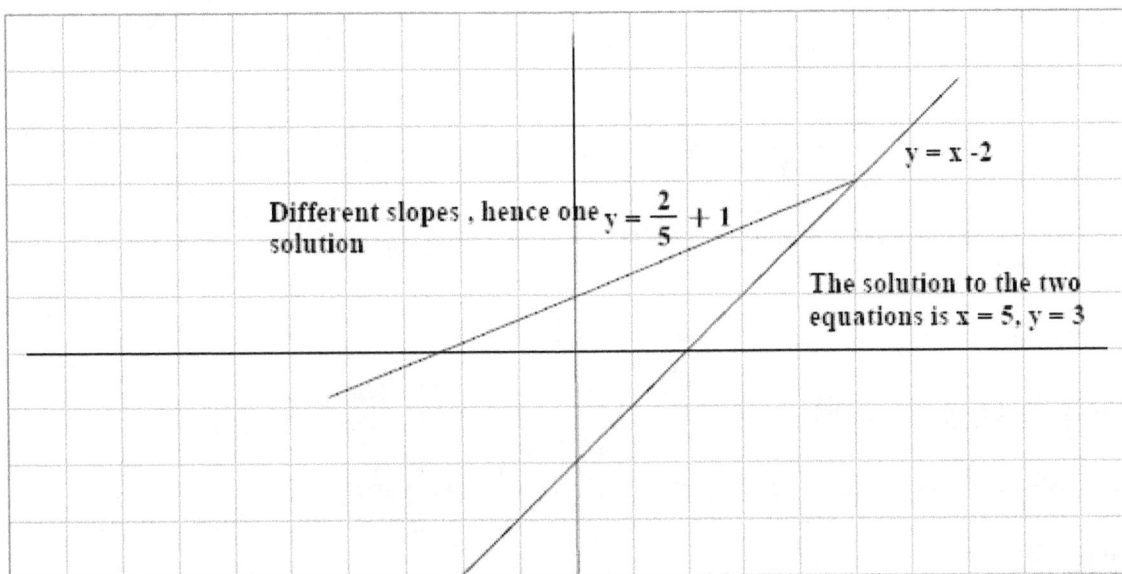

Different slopes, hence one solution $y = \frac{2}{5} + 1$

$y = x - 2$

The solution to the two equations is $x = 5$, $y = 3$

If two equations have the same slope and different **y** intercepts, then the two lines will never meet and there is no solution.

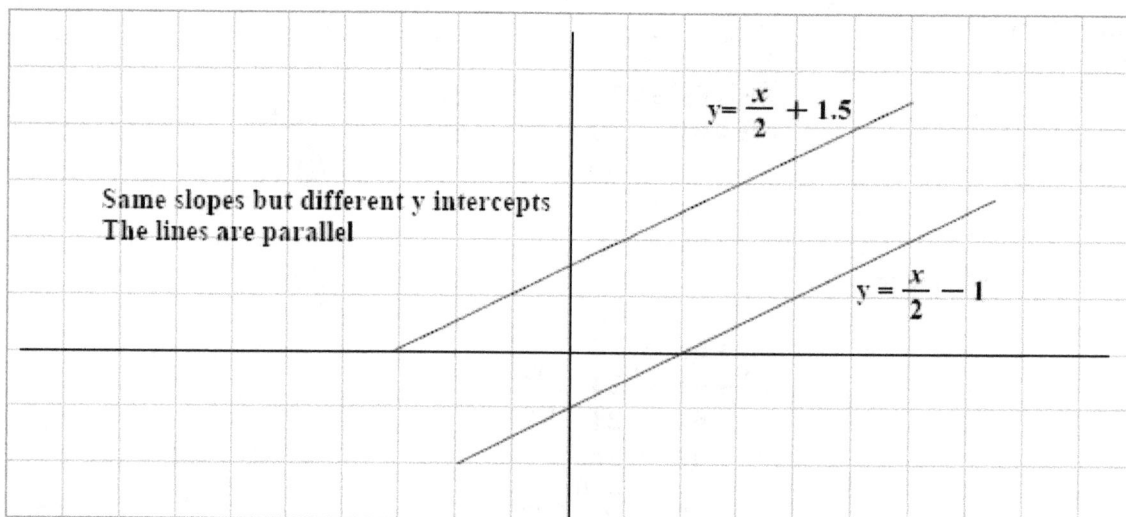

Same slopes but different y intercepts
The lines are parallel

$$y = \frac{x}{2} + 1.5$$

$$y = \frac{x}{2} - 1$$

Practice 2
Solve by elimination

a) $2x - 4y = 6$
$ 3x + 5y = 8$

b) $4x + 3y = 10$
$ 5x - 2y = 6$

c) $5x - 6y = 5$
$ 3x - 4y = 9$

d) $\dfrac{x}{4} - \dfrac{y}{3} = 2$

$ 2x + 3y = 8$

e $\dfrac{x + y}{6} = 4$

$ \dfrac{2x - 3y}{7} = 8$

f) $- 0.5x + 0.75y = 4$ (Hint Multiply by **100**)
$ 0.35x - 0.6y = 6$

Word Problems

Putting words into equations

Put the following words into algebraic expressions with two variables.

Example 5

Rembrandt and Picasso sold a total of **258** paintings. Rembrandt sold **25** more than Picasso. Let *r* represent the number of paintings sold by Rembrandt and *p* represent the number sold by Picasso.

$$r + p = 258$$

Rembrandt sold 25 more than Picasso $\quad r = p + 25$

Notice **25** is on the right hand side and not the left.

Looking at a numerical example.

$$9 \text{ is } 5 \text{ more than } 4$$
$$9 = 4 + 5 \qquad (5 \text{ is added to } 4)$$

Example 6

The length of a rectangle is **6** cm more than the width.

$$L = W + 6$$

If the same question is worded differently:- The width of a rectangle is **6** less than the length.

$$W = L - 6$$

Both equations are the same.

Example 7

Bob is three times as old as Mary.

$$b = 3m$$

Example 8

Sam is twice as old as Julie was seven years ago.

If Julie's current age is j, then seven years ago Julie was j-7

$$s = 2(j - 7)$$

Practice 3. (Put your own variables)

a) Tom is twice as old as Harry was **7** years ago.
b) The length of a rectangle is equal to the width increased by **6**.
c) With the current the speed of a boat is **15** km/h.
d) Against the current the speed of a boat is **12** km/h.
e) In a marathon measured in minutes, Ali finished **12** minutes before Jay.
f) Five times a number subtracted from **7**.
g) **8** subtracted from **6** times a number.

Example 1

A coffee store has two grades of coffee. One sells for $6 a kg and the other sells for $8/kg. The manager wants to mix the two grades to get **150** kg of coffee to sell for $7.5/kg. How much of each grade of coffee should be mixed?

There are two equations, one deals with the number of kilograms and the other deals with cost. Let the number of kilograms of the $6/kg coffee be x and the number of kilograms of the $8/kg coffee be y.

$$\text{then } x + y = 150 \qquad (1)$$

To find the cost of x kilograms of coffee at $6/kg, multiply the number of kilograms by the rate per kg.

$$\text{The cost of the \$6/kg coffee} = 6x$$
$$\text{Similarly the cost of the \textbf{\$8/kg} coffee} = 8y$$
$$\text{The total cost of \textbf{150} kg mixture at \$7.5/kg} = 150(7.5) = \$1125$$

Equating the cost of the two brands of coffee to the total cost.

$$6x + 8y = 1125 \qquad (2)$$
$$\text{From equation } \mathbf{1}, y = 150 - x$$

Substituting for y in equation **2**

$$6x + 8(150 - x) = 1125$$
$$6x + 1200 - 8x = 1125$$
$$-2x = -1200 + 1125$$
$$-2x = -75$$
$$x = 37.5$$

Substituting for x in equation **1**

$$37.5 + y = 150$$
$$y = 150 - 37.5$$
$$y = 112.5$$

The manager mixes **37.5** kg of the $6/kg coffee and **112.5** kg of the $8/kg coffee.

Example 2

John has **$230** in rolls of nickels and quarters. There are **40** nickels in a roll and **40** quarters in a roll. If the total number of rolls is **55**, find the number of rolls of each coin.

$$\text{One roll of nickels is worth } \mathbf{40} \times \mathbf{0.05} = \$2.$$
$$\text{One roll of quarters is worth } \mathbf{40} \times \mathbf{0.25} = \$10$$

Let the number of rolls of nickels be n and the number of rolls of quarters be q

$$n + q = 55 \qquad (1)$$
$$2n + 10q = 230 \qquad (2) \text{ (Since each nickel roll is \$2, to}$$

get the dollar value multiply the number of rolls by the dollar value of each roll)

$$\text{From Equation 1, } n = 55 - q$$

Substituting for n in Equation 2

$$2(55 - q) + 10q = 230$$
$$110 - 2q + 10q = 230$$
$$8q = 230 - 110$$
$$8q = 120$$
$$q = \frac{120}{8}$$
$$q = 15$$

Substituting for q in equation 1

$$n + 15 = 55$$
$$n = 55 - 15$$
$$n = 40$$

There are **40** nickel rolls and **15** rolls of quarters.

Interest

To calculate the interest on an investment, multiply the interest rate by the amount of the investment. The interest on **\$2000** for one year, if the interest rate is **6%** a year (per annum)

$$\text{is } 2000 \times \frac{6}{100} = \$120 \text{ per annum.}$$

Example 3

On a total investment of **10000**, part of it was invested at **8%** a year and the remaining at **10%** per annum and the income or interest from these investments totaled **\$900** per annum, how much of each investment was invested?

If the amount invested at **8%** is x dollars and the amount invested at **10%** is y dollars,

$$\text{then } x + y = 10000 \qquad (1)$$

Interest on the **8%** investment is $0.08(x)$ $(\frac{8}{100} = 0.08)$

Interest on the **10%** investment is $0.1(y)$

Equating the total interest of **\$900** to the interest of the **8%** investment and the **10%** investment.

$$0.08x + 0.1y = 900 \qquad \text{Multiply by } 100 \text{ to eliminate the decimals.}$$

$$8x + 10y = 90000 \qquad (2)$$

From equation **1**, y $=$ **10000** - *x*

Substituting for *y* in equation **2**

$$8x + 10(10000 - x) = 90000$$
$$8x + 100000 - 10x = 90000$$
$$-2x = 90000 - 100000$$
$$-2x = -10000$$
$$x = 5000$$
$$y = 10000 - 5000$$
$$y = 5000$$

5000 is invested at **8%** and **5000** is invested at **10%**

Example 4

Using ratios to solve percent problems.

The yield on Johnson and Johnson's stock is **3.5%** and the stock is trading at **$60**. What would the yield be if the stock dropped to **$45**?

Some stocks pay a dividend which is a fixed amount of money paid to the owner of the stock and does not change with the value of the stock.

Since the yield is **$3.5** on **100**,

the dividend on a **$60** stock would be $\dfrac{3.5}{100} = \dfrac{d}{60}$

$$d = \dfrac{60}{100}x3.5 \quad (d = \text{dividend})$$
$$d = 2.1$$

The dividend for Johnson and Johnson is **$2.1**. This dividend is fixed, so if the price of the stock dropped to **$45**, the dividend would still be **$2.1**.

To find out the yield or percent use ratios.

$$\dfrac{2.1}{45} = \dfrac{y}{100}$$
$$y = \dfrac{100}{45}x2.1 = 4.66 \text{ or } 4.66\%$$
$$\left(\dfrac{\textbf{part}}{\textbf{whole}} = \dfrac{\textbf{part}}{\textbf{whole}} \right)$$

The yield on Johnson and Johnson is **4.66%** when the stock trades at **$45**

Speed

$$\textbf{Speed} \;=\; \frac{\textbf{distance}}{\textbf{time}}$$

An easy way to remember this formula is to think of the speed of a car (kilometers per hour

or $\dfrac{\textbf{km}}{\textbf{hours}} \;=\; \dfrac{\textbf{distance}}{\textbf{time}}$)

By cross multiplying we can express time in terms of distance and speed; and distance in terms of speed and time.

$$s \;=\; \frac{d}{t} \qquad (s = \text{speed};\ d = \text{distance};\ t = \text{time})$$

$$\frac{s}{1} = \frac{d}{t}$$

$st \;=\; 1d$ (*The numerator of the first fraction is multiplied by the denominator of the second followed by an equal sign and then the denominator of the first fraction is multiplied by the numerator of the second fraction ; this process is cross multiplication*)

$st = d$

$t \;=\; \dfrac{d}{s}$ (*expressing in terms of* t)

$s \;=\; \dfrac{d}{t}$; $d = st$; $t = \dfrac{d}{s}$ (*remember the first formula and derive the other two*)

Example 5

A patrol boat take **4** hours to travel **50** km upsteam and **3** hours to travel the same distance downstream. Find the speed of the boat and the current.

Let the speed of the boat be b km/h and the speed of the current be c km/h
With the current the combined speed of the boat and the current is $b + c$
Against the current the combined speed is $b - c$

$$s \;=\; \frac{d}{t}$$

Against the current, $b - c \;=\; \dfrac{50}{4}$ ($\dfrac{50}{4}$ is the speed; distance divided by time)

$$b - c \;=\; 12.5 \quad (1)$$

With the current, $b + c \;=\; \dfrac{50}{3}$

$b + c \;=\; 16.66 \quad (2)$ (*Since c has opposite signs in the two equations there is no need to subtract or change signs*)

Rewriting equations **1** and **2** and then adding.

$$b - c = 12.5 \qquad (1)$$
$$b + c = 16.66 \qquad (2)$$
$$\text{-------------------}$$
$$2b = 29.16$$
$$b = 14.58$$

Substituting for **b** in Equation 2

$$14.58 + c = 16.66$$
$$c = 16.66 - 14.58$$
$$c = 2.08$$

The speed of the boat is **14.58** km/h and the current speed is **2.08** km/h

Acid solutions

If a **100** ml solution had an acid content of **60%**, the amount of acid would equal **60** ml (**60%** of **100** ml)

Example 6

What volume in millimeters of a **50%** acid solution must be added to **100** ml of a **30%** acid solution to make a **40%** acid solution?

Let the volume of the **50%** solution to be added equal x ml
The acid content of the **50%** solution is **0.5x** ml   (**50%**)(x)
The acid content of the **30%** solution is **0.3(100)** = **30**ml
The acid content of the final **40%** acid solution is **0.4(100 + x)** (*The total volume of the final solution is 100ml of the 30% solution plus the x ml of the 50% solution*)
Equating the acid content of the **50%** solution plus the acid content of the **30%** solution to the acid content of the final solution.

$$0.5x + 30 = 0.4(100 + x)$$
$$0.5x + 30 = 40 + 0.4x$$
$$0.5x - 0.4x = 40 - 30$$
$$0.1x = 10$$
$$x = 100$$

100ml of the **50** % solution must be added.

Example 7

A chemist has **100** ml of a **60%** acid solution and wants to reduce the strength to a **40%** acid solution by adding distilled water. How much water should be added?

The acid content of the **60%** solution is **60**ml, which stays the same.
Let w ml be the amount of water to be added.
(The acid volume of **60** ml stays the same. To the original amount of **100**ml, w ml of water is

added. To represent the acid content as a fraction, <u>the acid volume is the numerator and the total volume is the denominator</u>. This fraction must be equal to $\dfrac{40}{100}$

$$\dfrac{60}{100+w} = \dfrac{40}{100}$$

$$\text{Cross multiplying } 60(100) = (100+w)(40)$$
$$6000 = 4000 + 40w$$
$$6000 - 4000 = 40w$$
$$2000 = 40w$$
$$w = \dfrac{2000}{40}$$
$$w = 50$$

50 ml of water is added.

<u>Number representation</u>

Any two digit number has a units and a tens place.

Example

In the number **74**, **7** is in the tens place and **4** is in the units place.
The actual value of the number can be represented by $7(10) + 4(1) = 70 + 4 = 74$
If the digit in the tens place is represented by x and the digit in the units place is represented by y, then the actual number is $10x + y$

Example 8

The sum of the digits of a number is **7**. When the digits are reversed, the new number is **9** more than the original number. Find the number.

Let the digit in the tens place be x and the digit in the units place be y. The sum of the digits is **7**
$$x + y = 7 \quad (1)$$
The original number is $10x + y$
When the digits are reversed, the new number is $10y + x$
Since the new number is **9** more than the original number, **9** has to be added to the original number to make the equation true.(eg. If **8** is **3** more than **5**; add **3** to **5** and equate to **8**; **5 + 3 = 8**)
$$10x + y + 9 = 10y + x$$
$$10x - x + y - 10y = -9$$
$$9x - 9y = -9 \quad (2)$$
$$\text{From Equation 1, } y = 7 - x;\ \textit{substitute this value of } y \textit{ in}$$
$$\textit{equation 2}$$
$$9x - 9(7 - x) = -9$$
$$9x - 63 + 9x = -9$$
$$18x = 54$$
$$x = 3$$

Substituting this value of x in equation 1

$$3 + y = 7$$
$$y = 4$$

The number is **34**

When the digits are reversed, the new number is **43**, which is **9** more than **34**.

Example 9

The sum of two numbers is **350**. Their difference is **170**. Find the numbers.

In this example, digits are not important as reference is made only to the numbers.
Let the larger number be x and the smaller number be y

$$x + y = 350 \qquad (1)$$
$$x - y = 170 \qquad (2)$$

Add Equations 1 and 2

$$2x = 520$$
$$x = 260 \qquad \text{Substituting this value of } x \text{ in Equation 1}$$
$$260 + y = 350$$
$$y = 350 - 260$$
$$y = 90$$

The numbers are **260** and **90**

Example 10

Five times the smaller of two numbers plus seven times the larger is **74**. When five times the smaller is subtracted from seven times the larger, the result is **24**. Find the numbers.

Let the larger number be x and the smaller number be y.

$$5y + 7x = 74 \qquad \textit{(Five times the smaller plus seven times the larger} = \textbf{74})$$

Rearranging the equation above;

$$7x + 5y = 74 \qquad (1)$$
$$7x - 5y = 24 \qquad (2) \textit{ (five times the smaller is subtracted from seven times the larger and the result is } \textbf{24})$$

Adding Equations 1 and 2

$$14x = 98$$
$$x = 7$$

Substituting for x in equation **1**

$$7(7) + 5y = 74$$
$$5y = 74 - 49$$
$$5y = 25$$
$$y = 5$$

The smaller number is **5** and the larger number is **7**

Ratios

Example 11

There were **3** boys and **5** girls when the school opened. When a school bus arrived late with an equal number of boys and girls, the ratio of boys to girls was **8:9**. How many girls and boys arrived late?

Let the number of boys and girls that arrived late be x.

$$\frac{3+x}{5+x} = \frac{8}{9}$$ *(When **x** is added to the number of boys and*

*x is added to the number of girls, the ratio is **8:9**)*

Cross multiplying

$$9(3+x) = 8(5+x)$$
$$27 + 9x = 40 + 8x$$
$$9x - 8x = 40 - 27$$
$$x = 13$$

13 boys and **13** girls arrived late.

Example 12

Sally's and Bob's age are in the ratio of **9:10**. If the sum of their ages is **95**, how old is Sally and Bob?

Let Sally's age be *s* and Bob's age be *b*

$$\frac{s}{b} = \frac{9}{10}$$ *(The ratio of Sally's age to Bob's age is **9:10**)*

$$s = \frac{9b}{10} \quad (1) \quad$$ *(Moving **b** to the RHS of the*

equation)

The sum of their ages is **95**.

$$b + s = 95$$
$$s = 95 - b \quad (2)$$

Substituting the value of *s* from Equation 1 into Equation 2. (Replace *s* in Equation 2 with $\frac{9b}{10}$ from Equation 1)

$$\frac{9b}{10} = 95 - b \quad \textit{Multiplying both sides by **10**.}$$
$$9b = 950 - 10b$$
$$19b = 950$$
$$b = 50$$
$$s = 95 - 50$$
$$s = 45; b = 50$$

Sally is **45** and Bob is **50** years old.

Practice

1. Jacob's car wash charges $5 to wash a car and $7 a van. If **45** vehicles were washed and the total revenue was **$265**, how many of each type of vehicle were washed?

2. Investment **A** pays **4%** interest per annum and investment **B** pays **3.5%** per annum. If a total of **10 000** was invested and the total interest earned was **$380**, how much of each investment was invested?

3. One type of mixed nuts has **30%** cashews and another has a cashew content of **45%**. How many grams of each type of mixture must be mixed if **500** grams of a **39%** cashew content is required?

4. A lawn fertilizer is **24%** nitrogen and another fertilizer has a content of **18%** nitrogen. How much of each should be mixed to get a mixture of **100** kg that is **21%** nitrogen.

5. Angela made **10** litres of orange juice using **60%** water. How much orange concentrate must be added to the water to bring the water concentration down to **50%**. (Hint the **6** litres of water remain the same; let *x* litres = amt of concentrate added)

6. A plane flew a distance of **4000** km with the wind and took **5** hours. On the return flight against the wind, the plane took **6** hours and **40** minutes. Find the speed of the plane and the wind speed.

7. A van travels **70** km in the same time it takes a plane to travel **700** km. The speed of the plane is **630** km faster than the van. What is the speed of the van and the plane?

8. Sandra left Ottawa at **1** pm and travelled at **70** km/h. Jacob left Ottawa at **2**pm and travelled at **100** km/h. How far from Ottawa did they meet?

9. The yield on the Bank of Nova Scotia stock is **6%** and the stock is trading at **$50**. What is the yield when the stock drops to **$40**?

10. The sum of the digits of a number is **9**. When the digits are reversed, the new number is **9** less than the original number. Find the original number.

11. The sum of two numbers is **320** and their difference is **80**. Find the numbers.

12. The sum of two numbers is **105**. Three times the smaller plus twice the larger is **245**. What are the numbers?

13. Jim has three times as many $2 bills as $5 bills. If the total value of the bills are **$66**, how many of each type of bill does he have?

14. In a vending machine, the number of quarters is **5** more than twice the number of dimes. If the total value of the coins is **$25.25**, what is the number of quarters and dimes?

15. The sum of Jack's age and his father's age is **65**. Three times Jack's age increased by **5** equals his father's age. How old is Jack?

16. At a party there were **30** men and **25** women. After the party started, twice as many men as women arrived late. The ratio of men to women was now at **10:7**. How many men and women arrived late?

17. John has **$13** in change consisting of dimes and quarters in the ratio of **3:4**. How many dimes and quarters does he have?

18. Malik drove a distance of **500** km in **5** hours. He drove part of the distance at **90** km/h and part of the distance at **105** km/h. What distance did he drive at each speed?

Answers
CHAPTER 1
INTEGERS

Practice 1. p. 5
a) -5   b) 31   c) 20   d) 66   e) 79   f) 0   g) 9   h) -6
Practice 2 and 3   p. 6
a) 576   b) 320   c) 1120   d) -504   e) -360
3. a) -5   b) -3   c) 4   d) 5   e) -4   f) -5

CHAPTER 2
FRACTIONS DECIMALS AND PERCENT

Practice 1and 2   p. 9
a) M: $\frac{1}{4}$; J: $\frac{1}{3}$; $\frac{7}{12}$ b) C: $\frac{1}{6}$; A : $\frac{5}{6}$ c) m: $\frac{1}{5}$; r: $\frac{4}{5}$ d) r: $\frac{1}{6}$; g : $\frac{1}{3}$; y: $\frac{1}{2}$

2. a) $\frac{3}{4}$ b) $\frac{8}{9}$ c) $\frac{5}{6}$ d) $\frac{5}{6}$ e) $\frac{2}{3}$ f) $\frac{3}{7}$ g) $\frac{9}{17}$ h) $\frac{17}{18}$

Practice 3   p. 12
a) 3,2,2,7,11   b) 3,7,47   c) 3,5,23   d) 3,3,3,3,3,3   e) ten 2's   f) 5, 251   g) 2, 391
Practice 4.   p.14
a) 360   b) 24   c) 364   d) 420   e) 315   f) 90   g) 90   h) 12
Practice 5.   p. 15
a) $\frac{17}{20}$ b) $-\frac{13}{63}$ c) $-\frac{1}{12}$ d) 0   e) $\frac{4}{15}$ f) $1\frac{5}{12}$ g) $-\frac{1}{8}$

Practice 6.   p.16
a) $\frac{3}{4}$ b) $\frac{7}{8}$ c) $\frac{5}{8}$ d) $\frac{3}{5}$

Practice 7.   p.17
a) $\frac{9}{128}$ b) $\frac{5}{8}$ c) $\frac{9}{23}$ d) $\frac{16}{19}$ e) $\frac{46}{73}$ f) $\frac{1}{5}$ g) $\frac{1}{3}$ h) $\frac{13}{29}$ i) $\frac{11}{37}$

Mixed numbers

Practice 1.   p.18
a) $7\frac{1}{6}$ b) $7\frac{2}{5}$ c) $8\frac{1}{3}$ d) $13\frac{2}{5}$ e) $3\frac{6}{7}$ f) $7\frac{7}{8}$ g) $9\frac{2}{3}$ h) $7\frac{2}{7}$

Practice 2.   p. 19
a) $\frac{19}{4}$ b) $\frac{29}{8}$ c) $\frac{47}{8}$ d) $\frac{41}{7}$ e) $\frac{27}{4}$ f) $\frac{69}{8}$ g) $\frac{63}{8}$ h) $\frac{23}{3}$ i) $\frac{38}{7}$ j) $\frac{49}{5}$ k) $\frac{47}{4}$

Practice 3.   p. 20
a) $\frac{1}{4}$ b) $\frac{15}{28}$ c) $\frac{6}{7}$ d) $\frac{42}{55}$ e) $\frac{18}{35}$ f) $\frac{143}{119}$ g) $\frac{6}{11}$ h) $\frac{17}{15}$

Decimals

Practice 1. p. 21
a) 0.4   b) 0.8   c) 0.3   d) 0.5   e) 0.3
Practice 2. a) 0.75   b) 0.28   c) 0.55   d) 0.39   e) 0.15
Practice 3. a) 0.875   b) 0.874   c) 0.589   d) 0.357   e) 0.563
Practice 4. p. 23
a) 16.002   b) 45.56   c) 8784.3   d) 2295   e) 5013.88   f) 1050   g) 2562.5

Decimal form

Practice 1 p. 24
 a) 0.000001   b) 0.000000001   c) 0.001   d) 0.0000001
2. a) 100000   b) 1000000000   c) 10000   d) 10000000   e) 1000
3. a) 10^3 b) 10^4 c) 10^7
4. a) 67800   b) 0.873   c) 9.780   d) 0.00000768   e) 0.000234   f) 7   g) 6   h) 7.8   i) 0.678   j) 32.4

Scientific notation

a) 4.76×10^8 b) 5.96×10^6 c) 9.47×10^8 d) 8.34×10^9 e) 7.8×10^{-8}
f) 7.86×10^{-7} g) 7.6×10^{-6}
6. a) 30541.5   b) 378.314   c) 50.6321   d) 91851.45
Practice 5. p. 25
7. p. 27

F	D	P
$\frac{11}{20}$	0.55	55%
$\frac{2}{5}$	0.4	40%
$\frac{7}{8}$	0.875	87.5%
$\frac{9}{20}$	0.45	45%
$\frac{3}{8}$	0.375	37.5%
$\frac{8}{25}$	0.32	32%
$\frac{23}{50}$	0.46	46%

Practice 8. p. 28
 a) 28560   b) 48.88   c) 150 @ 15%   d) 25,000   e) 878.77   f) 3.64   g) 15243.90

CHAPTER 3
RATES AND RATIOS

Practice 1. p. 30
 a) 10   b) 8.4   c) 9   d) 3.5   e) 12   f) 3.375   g) 13.33
Practice 2. p. 35
a) c:2; s:3; v:4   b) o:9; a:21   c) t:6, r:15   d)1. o:16; n:24; b:32   2. o:40; n:60; b:80   3.
o:30; n:45; b:60   4. o:36; n:54 b:72 . e) 600   f) r:30; g:40; y:50   g)120
3. a)166.67   b) 6.4   c) 75%   d) 18   e) 600
4. a) 30.30   b) 4.65   c) 3.95   d) 26.35
Rates
 Practice 1. pp. 36,37
1. a) 100 words per minute   b)7.27 m/s   c) 75 cents per metre   d) andrew   e) 1.33 cents/gm
 f) 71.43c/litre

2. a) 857   b) 3500   c) 1800   d) $30
3. Hilton
4. Brand A
5. 350 km
6. 4166 km

CHAPTER 4
MEASUREMENTS

Practice p. 40
1. a) 5890 m   b) 0.056m   c) 4.73m   d) 6.75m   e) 8.954m   f) 75.93m
2. a) 3500cm   b) 57cm   c) 4.75×10^5cm   d) 0.263cm   e) 3850cm   f) 0.834cm
3. a) 0.35m   b) 3750m   c) 3.45×10^{-6}km   d) 6.4mm   e) 14.7×10^5cm   f) 6340m
Practice p. 42
1. a) 4.4×10^{12}mm^2 b) 6.43×10^6m^2 c) 5400mm^2
2. a) 5.89×10^{-4}m^2 b) 0.055m^2 c) 5.44×10^7m^2
3. a) 0.5cm^2 b) 4.5×10^{11}cm^2 c) 85×10^4cm^2
                                                                    Volume
Practice p. 45
1. a) 9.6×10^4 litres   b) 5×10^{-12}km^3 c) 5×10^{-12}km^3 d) 2×10^{18}km^3
2. a) 6   b) 8   c) 9   d) 10   e) 5   f) 3   g) 4
3. a) 2   b) 4   c) 5   d) 7   e) 9

CHAPTER 5
Equations 1

Practice p. 46
1. a) x=32 b) x=12 c) y= -5 d) z=49 e) t=49 f) c=49 g) h=101 h) k=1
Practice p. 50
2. a) $r = \dfrac{c}{6}$ b) $r = \dfrac{v}{6h}$; $h = \dfrac{v}{6r}$ c) $W = \dfrac{A}{l}$; $l = \dfrac{A}{w}$ d) $l = \dfrac{V}{wh}$; $w = \dfrac{V}{lh}$; $h = \dfrac{V}{wl}$
e) $b = \dfrac{2a}{h}$; $h = \dfrac{2a}{b}$ f) $y = \dfrac{-25x}{6} + 16\dfrac{2}{3}$; $x = \dfrac{-6y}{25} + 4$ g) sum of parallel sides= $\dfrac{2A}{h}$
3. a) 5 b) 5 c) 4 d) 36 e) -20 f) -21 g) -25 h) -12 i) -8

Patterns
Practice p. 52
1. a) 2n - 1 ; 59 b) 3n + 7 ; 97 c) 4n + 11 ; 131 d) 7n + 18 ; 228
2. 7
3. a) 12 b) 12 c) 11
Practice p. 55
4. a) $n^2 + 1$ b) n(n+1) c) n(n+1) +2 d) $(n + 1)^2$ e) $(n + 1)^2 + 2$ f) $(n + 3)^2$
Mean mode median
Practice p. 57
1. a) 50.375, 54, 54 b) 53.75, 55, 56 c) 50.375, 34, 50.5 d) 32, 24, 29 2. a) 65 b) 25 c) 20 and 30
3. no answer

CHAPTER 6
GEOMETRY 1

Practice P. 60
1. a) square, rhombus b) square, rectangle c) square, rhombus and kite d) parallelogram, rhombus and kite
Practice p 62
2. a) $10cm^2$ b) $42cm^2$ c) $27cm^2$
Practice p. 63
3. 10cm 4. 15cm 5. 15cm 6. 40cm
Practice pp. 67, 68
7) 44 cm 8) $78.5cm^2$ 9) $377cm^2$ 10) $1885 cm^3$ 11) 8cm 12) 58.31m
13) 10cm
Practice p. 70
14 a) $214 cm^2$ b) $348 cm^2$ c) $334cm^2$

Practice pp. 73, 74, 75
1. a) 10cm^2 b) 30cm^2 c) 13.5cm^2 d) 31.42cm^2 e) 282.74cm^2
2. A: 12.86cm^2 B: 44 cm^2 C : 44.27 cm^2 D : 1.93 cm^2
3. a) ABC = 5 cm^2 , 4 cm^2, 4 cm^2, 4 cm^2 b) $\frac{1}{4}$

4) 10cm 5) 6cm 6) 16cm 7) 6cm 8. a) SA =226.16cm^2; V=251.33cm^3
8. b)SA=94 cm^2; V = 60 cm^3
8. c) A: SA=46.3 cm^2; V = 10 cm^3 B: SA=240 cm^2; V = 252 cm^3
9) 706.86cm^2 10) 3cm   11) 3cm

12. a) 25 cm^2 b) 37.5 cm^3 c) 50 cm^3 d) 150 cm^3 e) 6.25 cm^3
Angles
Practice pp. 79, 80
1. a) 110°; b=c=70° b) Z: a and c; b and d C: a and d; b and c
c) A: x=60°, y=30° B: a=70°, b=60°, c=50°, d=60° C: a=58°, b=c=70° D: a=60°, b=c=60°
d) b=c=65°
2. A: a=70°, b=110°, c=d=60°, e=50°, f=70°
B: a=c=130°, b=50°,d=e=140°, f=40°, g= j = 90°, h= i =90°, k=40°, n=50°, m=130°
3. B = D =125°

CHAPTER 7
BEDMAS 1

Practice p. 83
1. a) 7 b) - 1 c) 112 d) 10 e) -375 f) 30 g) 28 h) 2 i) 53 j) 4 k) -50868 l) - 5472
2. a) -4 b) 5 c) 6 d) -33
3. a) -11 b) -6 c) -56 d) -95

CHAPTER 8
TRANSFORMATIONS

Practice pp. 88, 89
1. A(-1, -2) B(-8, -3) C(-10, -4)
2. X(10, 7) Y(-1, 8) Z(7, 0)
3. K'(4, 5) L'(2, 4) M'(-4, -3)
4. P'(-3,- 5) Q'(3,- 4) R'(-2,- 6)
5. J'(0, 8) K'(-9, -1) L'(-3, 11)
6. A(-4, 9) B(0, 2) C(-8, 0)
7. W'(5, 2) X'(4, -7) Y'(-4, 6)
8. A'(3, -5) B'(-6, 4) C'(7, - 4)
9. J(-4, 6) K(3, -7) L(-5, 4)

CHAPTER 9
PROBABILITY

Practice pp. 93, 94, 95

1. a) $\frac{1}{24}$ b) $\frac{1}{12}$

2. a) $\frac{1}{8}$ 12.5%   b) $\frac{5}{16}$ 31.25%

3. a) $\frac{1}{13}$ b) $\frac{1}{26}$

4. a) $\frac{3}{16}$ b) $\frac{3}{8}$ c) $\frac{1}{16}$

5. a) $\frac{1}{2}$ b) $\frac{2}{3}$ c) $\frac{1}{6}$

CHAPTER 10
REVIEW OF ESSENTIALS SKILLS FOR GRADE 9

Practice pp. 99, 100, 101

1. a) $\frac{-14}{15}$ b) $\frac{31}{24} = 1\frac{7}{24}$ c) $-\frac{1}{8}$ d) 0.95   e) $2\frac{3}{11}$ f) $2\frac{3}{5}$

2. a) 6   b) 4.88%   c) \$103.19   d) 26.25   e) 18.75   f) 342.86

3) $3x + 10$ 4.) $0.25x + 1.25$ 5.) $x+6$ 6) $3x$ 7) x^2 8) $4-2x$ 9) $2x$ 10) $6-x$ 11) $x-5$

12) $\frac{x}{2} - 4$

CHAPTER 11
EXPONENTS

Practice pp. 104, 105

1. a) a^{12} b) a   c) 5^{14} d) 3^3 e) m

2. a) 2^2 b) 5^2 c) 6^6 d) 8^2

3. a) 3^{24} b) 8^{12} c) 10^{16}

4. a) 3   b) 1   c) $\frac{1}{2^2}$ d) $\frac{1}{3^{12}}$ e) 1   f) 2^7 g) 10^{10} h) $\frac{1}{c^2}$ i) 5^3 j) 1

5. a) $27a^9b^6x^9$ b) $8x^6b^9a^{15}$

Practice p. 106

6. a) Undefined   b) 0   c) 0   d) 1   e) 1   f) 1

Practice pp. 108, 109

7. a) 27   b) -64   c) -625   d) -125   e) x^3

8. a) $\frac{9x^8}{y^6}$ b) $\frac{b^8}{25y^6}$ c) $\frac{c^{18}}{64x^{12}}$ d) -8   e) $-\frac{x^{15}}{8y^9}$ f) $-\frac{x^{15}}{32}$ g) $81c^8$ h) $\frac{1}{x^{20}y^{28}}$

i) $\dfrac{y^{24}}{x^{12}}$ j) $\dfrac{16x^6}{25y^6}$ k) $\dfrac{729y^6}{512x^9}$ l) $\dfrac{y^3}{x^3}$ m) $\dfrac{4}{3}$

9. a) 665 b) 15 c) 9750000 d) $-64\dfrac{1}{125}$ e) 16 f) $16\dfrac{1}{80}$ g) $27\dfrac{2}{3}$ h) 3^6or 729 i) 3.81

 j) 49 k) 8 l) 1 m) 4096 n) -125 p) 256 q) -256

10. a) $\dfrac{9}{16}$ b) $\dfrac{64}{125}$ c) $\dfrac{64}{49}$

11. a) $4y^4$ b) $2y^{16}$ c) $5c^{14}$ d) $8x^{20}$

CHAPTER 12
BEDMAS 2

Practice p. 110

1) -292 2) $-\dfrac{32}{723}$ 3) 172 4) 272 5) -368 6) 659

7. a) 108 b) 18 c) 351 d) 0

8. a) -18 b) 2 c) 8 d) 4

CHAPTER 13
POLYNOMIALS

Practice p. 111

1. a) $-\dfrac{4}{3}$ b) $\dfrac{2}{5}$ c) -4 d) $-\dfrac{1}{7}$

Practice p. 112

2. a) $3x^2yz,\ 10x^2yz$ b) 5ab, 7ba, -6ab c) 3xz, 4zx, 5zx d) -4ab, 7ba; 3ac, 6ca

e) 4xy, -7yx, $\dfrac{yx}{2}$ f) 3 g) 8 h) 9

Practice p 113

3. a) m b) -11p c) 0.6x d) $-3x^3 + 13x^2$ e) $-\dfrac{8a}{15} + 3\dfrac{1}{4}b$ f) $13x^2$-xy +8 g) -2b +15

 h) -5x +7y i) $9x^2y - 6xy - 7x^2$ j) $\dfrac{m}{15} + 2\dfrac{14}{15}$n k) $-5\dfrac{1}{2}a - 6\dfrac{3}{4}b$

Practice p. 114

4. a) $x^2 + 2xy + y^2$ b) $x^3y + x^2y^2 + 2xy + 2y^2$ c) $4x^2y - 4x^2c + 4xcd$

Practice pp. 116, 117

5. a) 8x -2 b) x-5 c) x+3 d) $\dfrac{x}{40} + \dfrac{9}{16}$ e) $\dfrac{5c}{2} + 2.1$ f) 13.44s + 19.52t +7.36

6. a) 8b + 12c +56 b) -3x-3y c) 8x+11y d)2x - 9y +39z e)2s +4t

7. a) 2x + 3 b) 4x + 28 c) 4

8. a) -3 b) 0

9. $5\dfrac{2}{5}$

Practice pp. 117, 118

10. a) 31 b) 5 c) $-14\dfrac{2}{5}$

11. a) $-1\frac{7}{60}$ b) $2\frac{4}{7}$ c) $\frac{17}{72}$ d) $5\frac{8}{9}$ e) $1\frac{2}{3}$ f) $15\frac{1}{25}$ g) $-\frac{48}{49}$ h) $2\frac{5}{8}$ i) $2\frac{2}{7}$ j) $1\frac{13}{27}$

k) $4\frac{2}{9}$

12. a) $24x^4y^5$ b) $8a^4b^4c^2$ c) $14xyzabc$ d) $2a^2bx^4$

13. a) $\frac{1}{6xc}$ b) $\frac{10}{ab}$ c) $\frac{V^3}{4xya^3c^4}$ d) $\frac{a^{13}c^{14}}{x^{18}}$ e) $\frac{1}{64\,s^9t^{12}}$ f) $\frac{v^6}{32\,xa^4z^3}$

Polynomials chapter review

Practice pp. 119, 120

1. a) $-\frac{11a}{8}+\frac{7b}{10}$ b) $-\frac{m}{6}+\frac{3\,n}{4}-\frac{y}{8}$ c) $1.3p+8.7q$

2. a) $-2a-8b$ b) $-4x+7y-2z$ c) $k-10+3m$

3. a) $12a$ b) $20a+10c$ c) $ac+8ad+2bc+16bd$ d) $3a+6b+9c$ e) $-\frac{x}{2}-\frac{y}{12}+\frac{4z}{15}$

f) $\frac{4a}{15}-\frac{3b}{10}+\frac{6c}{25}$

4. a) $8m^3-14m^2+9m$ b) $9a^2-46a+19ab$

5. a) $5(2a+3)$ b) $5b(a-3c)$ c) $3ab(6c-7)$ d) $2x(2x^2+3x+4)$ e) $7x(2x^2+3x+4)$

   f) $2xy(x^3y^2-2x^2y+3)$

6. a) 27 b) $3\frac{6}{7}$ c) -60 d) 18.7

7. a) 8 b) -55 c) $2\frac{22}{27}$ d) 2.59 e) -16

8) $12x$ 9) $4x+2y+20$ 10) SA$= 12xy+18yz+12xz$; V$=18xyz$ 11) $14x+2y$

CHAPTER 14
EQUATIONS 2

Practice p. 126

1. a) $-\frac{3}{5}$ b) 2 c) -1 d) $-\frac{9}{7}$ e) $-\frac{7}{6}$ f) $-\frac{3}{4}$

Practice p. 128

2. a) A: linear   b) B: linear   c) C: not linear

Practice p. 130

4. a) $\frac{7}{4}$ b) $\frac{8}{5}$ c) -2 d) 4 e) -4 f) 5

Practice p. 131

5. a) neither   b) perpendicular   c) neither   d) parallel   e) neither   f) perpendicular.

Practice p. 132

6. a) $y=\frac{3x}{11}+6\frac{1}{11}$ b) $y=-\frac{x}{7}+4\frac{3}{7}$ c) $y=\frac{x}{2}-6$ d) $x=6$ e) $y=4$

Practice p. 133

7. a) $y = -2x + 10$ b) $y = \dfrac{x}{3} + 4\dfrac{1}{3}$ c) $y = \dfrac{-x}{5} + 4$ d) $x = 6$ e) $y = 8$

Practice p. 136

8. a) $x = \dfrac{24}{11} : y = \dfrac{10}{11}$ b) $x = \dfrac{10}{7} : y = \dfrac{6}{7}$ c) $x = -3: y = 6\dfrac{1}{2}$

Practice pp. 139, 140

1. a) $C = 0.5t + 8$ b) \$11   c) \$8

2. a) Parcel post $C = 1k + 4$: Courier express $C = 1.5k + 3$

c) At the point of intersection both services are the same (2kg)   d) Parcel post

3. a) At 100 km a day, both plans cost the same.   b) Plan A is better if you are driving more than 100 km a day.

c) Plan B

4. a) At 2 hours both charge the same   b) John

5. a) $C = 0.06c + 2$ b) $R = 0.09c$ c) At 67 cups, the revenue = cost.   6) 33.33 ml   7) 2 lbs

   8) 32 litres   9) \$5000 in each investment.   10) P:32 cm ; A: 39cm^2

CHAPTER 15
GEOMETRY 2

Practice pp. 145, 146

1. a) $a = 110°$; $b = 70°$; $c = 70°$ b) Z: a and c; b and d; C: b and c; a and d

2. $x = 83°$, $a = 75°$, $b = 75°$, $c = 30°$, $d = 110°$, $e = 60°$, $f = 50°$, $g = 44°$, $A = 88°$, $B = 44°$, $C = 48°$

3. AB=12.71cm, BC=8.14cm, AC=7.14cm; $x = 36°$, $a = 72°$, $b = 108°$, $c = 72°$, $d = 108°$, $e = 108°$, $f = 72°$, $g = 60°$, $y = 120°$, $z = 140°$

Practice p.151

4. a) $120°$,   b) $108°$ c) $144°$ d) $135°$

5. a) 192 cm^2 b) 75cm^2 c) 100cm^2

6. a) 628.3 cm^3 b) 640 cm^3 c) 1980cm^3

Practice pp. 153, 154

7. a) 392.7 cm^3 b) 130.9 cm^3

8.) 120   9) 65.97 cm^3 10) 4cm   11) 6 cm   12) 3 cm

Practice pp. 156, 157

13) 900% (9 times)   14) 144%

15. a) 300%   b) 800%   16) 800%   17) 25%

Chapter Review

Practice p. 159

1.) 33.51 cm^3 2) SA = 131.82 cm^2, V=88.8 cm^3 3.) 135   4.) 173.76 cm^2 5.) 173.2 cm^3

6.) Cone: 56.54cm^3; P:72cm^3 ; Pyramid larger by15.4cc   7.) 4cm   8.) 32.72cm   9.) 60°

Geometry of packaging

Practice pp. 161, 162

1.a) r = 4.57cm, h = 9.14 cm   b) 393.67 cm^3 c) 4.72c.

2. a) $r = \sqrt{\dfrac{A}{\Pi}}$ b) $r = \sqrt[3]{\dfrac{V}{8\,\Pi}}$ c) $r = \sqrt{\dfrac{SA}{6\,\Pi}}$

3.) r = 3.17 cm, h = 6.34 cm   4) r = 2.19 cm, h = 4.38 cm

Practice p. 163

5) 14.6 m^2 6) l = w = h = 51.74 cm

7) Cylinder = 3391 cm^3 ; Rectangular prism = 3007 cm^3, difference = 384 cm^3

8) l = w = h = 20 cm

CHAPTER 16
EQUATIONS 3

Practice p 166

1. a) $x = 1\frac{1}{7}, y = \frac{2}{7}$ b) $x = 5\frac{3}{5}, y = \frac{1}{5}$ c) x = -3, y = 4   d) $x = \frac{3}{2}, y = 1\frac{3}{4}$

Practice p. 169

2. a) $x = 2\frac{9}{11}, y = -\frac{1}{11}$ b) $x = 1\frac{15}{23}, y = 1\frac{3}{23}$ c) x = -17, y = -15   d) $x = 6\frac{2}{17}$,

$y = -1\frac{7}{17}$

e) $x = 25\frac{3}{5}, y = -1\frac{3}{5}$ f) $x = -184, y = -117\frac{1}{3}$

Practice p. 170

3. a) T = 2(h-7)   b) L = w + 6   c) b + c = 15   d) b - c = 12   e) A = J + 12   f) 7 - 5x   g) 6x – 8

Word problems

Practice pp. 179, 180

1. 25 cars, 20 vans   2) $4000 at 3.5% , $6000 at 4%   3) 200gms at 30%, 300gms at 45%

4) 50kg at 24%; 50kg at18%   5) 2 liters of orange   6) plane speed = 700km/h; wind speed = 100 km/h

7) Van speed = 70 km/h; plane speed = 700 km/h   8) 233.34 km   9) 7.5%   10)54

11) 200, 120   12) 35, 70

13) 6 $5, 18 $2   14) 40 dimes 85 quarters   15) Jack 15   16) 10 women 20 men

17) 30 Dimes, 40 quarters   18) 150 km at 90 km/h 350 km at 105 km/h

WWW.UNIVERSALMATH.NET

WWW.UNIVERSALMATH.NET

www.ingramcontent.com/pod-product-compliance
Lightning Source LLC
Chambersburg PA
CBHW051213200326
41519CB00025B/7098